Problems and Worked Examples in Chemistry

TO ADVANCED LEVEL

By the Same Authors

A NEW CERTIFICATE CHEMISTRY

GRADED PROBLEMS IN CHEMISTRY
TO ORDINARY LEVEL

CLASSBOOK OF PROBLEMS IN CHEMISTRY
TO ADVANCED LEVEL

WORKED EXAMPLES AND PROBLEMS IN
ORDINARY LEVEL CHEMISTRY

THE ESSENTIALS OF QUALITATIVE ANALYSIS

By A. Holderness

ADVANCED LEVEL INORGANIC CHEMISTRY

ADVANCED LEVEL PHYSICAL CHEMISTRY

ORDINARY LEVEL REVISION NOTES IN CHEMISTRY

By T. Muir and J. Lambert

PRACTICAL CHEMISTRY

Problems and Worked Examples in Chemistry
TO ADVANCED LEVEL

by

A. HOLDERNESS M.SC., F.R.I.C.

Formerly Senior Chemistry Master at Archbishop Holgate's Grammar School, York

and

JOHN LAMBERT M.SC.

Formerly Senior Chemistry Master at King Edward's School, Birmingham

THIRD EDITION

HEINEMANN EDUCATIONAL

Heinemann Educational Books Ltd
Halley Court, Jordan Hill, Oxford OX2 8EJ

OXFORD LONDON EDINBURGH MADRID
ATHENS BOLOGNA PARIS MELBOURNE
SYDNEY AUCKLAND SINGAPORE TOKYO
IBADAN NAIROBI HARARE GABORONE
PORTSMOUTH NH (USA)

ISBN 0 435 65438 1

© A. Holderness *and* J. Lambert 1959, 1972, 1978
First published 1959
Reprinted eight times
Second Edition 1972
Reprinted 1974, 1976
Third Edition 1978

95 96 97 98   14 13 12 11

Set in Monophoto Times 10/11
Printed in Great Britain by
Athenæum Press Ltd, Gateshead, Tyne & Wear

Contents

Chapter		Page
1	Acid–Alkali Titrations	1
2	Redox Titrations	12
3	Silver Nitrate Titrations	27
4	Gravimetric Analysis	32
5	Colligative Properties Not Involving Dissociation or Association	36
6	Colligative Properties Involving Dissociation or Association	49
7	Relative Atomic and Molecular Masses by Relative Vapour Density and Other Methods	54
8	Thermal Dissociation	63
9	Energetics	67
10	Distribution Constant	77
11	Electrolysis	82
12	Equilibria	85
13	Formulae of Hydrocarbons by Explosion with Oxygen	100
14	Formulae of Organic Compounds	103
15	Gas Analysis	111
16	Miscellaneous Problems	114
	Answers to Numerical Examples	125
	Logarithms	130
	Relative Atomic Masses and Other Useful Data	132

Preface

This book is intended for the use of students preparing for the Advanced Level examinations of the G.C.E. in Chemistry. The problems set for solution are identical with those in the same authors' *Class Book of Problems in Chemistry to Advanced Level* and are numbered in the same way, so that the two books may be used together. In addition, the present volume presents a brief statement of the essential theory connected with the various types of problem covered, as well as worked examples of each type. It is hoped that these additions will add to the value of the book.

Answers have been worked using four-figure logarithms and approximate relative atomic masses taken from the list supplied. We wish to thank the Senate of the University of Sheffield for permission to use certain Inter. B.Sc. questions. No problems set by examination boards are included. Advanced Level G.C.E. problems can, therefore, be used as additional unseen tests if desired.

Sample calculations involving oxidation and reduction are solved in terms of electron gain and electron loss. It is hoped that teachers will welcome this approach, which is in line with the newer presentation of oxidation and reduction.

A. H.
J. L.

November, 1958

Preface to the Second Edition

This edition has been prepared with the new Advanced Level Syllabuses in mind, particularly those of the Joint Matriculation Board and the University of London Examinations Board. Some chapters have been regrouped and the titles have been brought up to date. New worked examples and problems have been included and others, which are no longer relevant, have been omitted. The methods of solving problems have been modified where it was thought that a more fundamental approach would be in keeping with modern methods of teaching chemistry. The revision has been most extensive in the chapters on Energetics and on Titrimetric Analysis.

SI units have been used except where a traditional unit is likely to remain in use for some time. For example, the convenient mmHg is included as an alternative for Nm^{-2} as a unit of pressure. Normalities have been abandoned completely and molarities are used throughout.

In the system of nomenclature employed, an attempt has been made to cover the present transition period. Compounds containing metals are named entirely by the Stock notation. Organic compounds are either named systematically or, if it is thought that the traditional name is still in use, but is likely to be replaced, both are given.

January, 1972
J. N. L.

Preface to the Third Edition

Although the chapters are presented in the same order as for the Second Edition and no new problems or worked examples have been included, opportunity has been taken to bring the nomenclature and terminology used in the book completely in line with current recommendations and to correct any outstanding errors in both text and answers.

My thanks are due to Mr J. S. Clarke of Alleyn's School for assistance in preparation of this edition.

September, 1978
J. N. L.

1

Acid–Alkali Titrations

Theory

Titrimetric Analysis is a method of quantitative analysis which involves the experimental determination of volumes of solutions which react together completely.

Concentrations of the solutions are usually measured in either grams of solute per dm^3 of solution ($g\,dm^{-3}$) or moles of solute per dm^3 of solution ($mol\,dm^{-3}$).

The concentration of a solution expressed in $mol\,dm^{-3}$ is known as the molarity of the solution.

Consider sodium hydroxide of formula NaOH.

$$1 \text{ mole of NaOH} = 23 + 16 + 1 = 40\,g$$

A solution containing 40 g of NaOH in 1000 cm^3 of solution is known as a MOLAR or 1M solution.

A solution containing 20 g of NaOH (half a mole) in 1000 cm^3 of solution is known as a half molar or 0.5M solution.

A solution containing 4 g of NaOH per 1000 cm^3 of solution is known as a decimolar or 0.1M solution.

The two relationships which are required when using the molarity system are:

Number of moles = $\dfrac{\text{mass of substance in g}}{\text{mass of 1 mole of substance in g}}$

Molarity = number of moles of substance in one dm^3 of solution

When using a titration for determining the concentration of a solution it is necessary to know the balanced equation for the reaction. The balanced equation gives the ratio of the numbers of moles of substances reacting together. For example,

$$2NaOH + H_2SO_4 \rightarrow Na_2SO_4 + 2H_2O$$

This means that, whenever a solution of sodium hydroxide reacts completely with a solution of sulphuric acid, the number of moles of NaOH will always be equal to twice the number of moles of H_2SO_4. It is the mole ratio, derived from the equation, which is the basis for all calculations.

ACIDS AND ALKALIS

Reactions between acids and alkalis are particularly suitable for use in Titrimetric Analysis as they are rapid, essentially complete and the point at which the reaction is complete is easily detected.

All reactions between acids and alkalis can be represented by:

$$H^+ + OH^- \rightarrow H_2O \quad \text{or} \quad H^+(aq) + OH^-(aq) \rightarrow H_2O(l)$$

That is 1 mole of H^+ ions always combines with 1 mole of OH^- ions. If an acid containing 2 moles of H^+ ions per mole of acid (e.g., H_2SO_4) reacts with an alkali containing 1 mole of OH^- ions per mole of alkali (e.g., NaOH), then 2 moles of the alkali would be needed to react completely with 1 mole of acid. This same result can equally well be obtained from the full stoichiometric equation.

$$2NaOH + H_2SO_4 \rightarrow Na_2SO_4 + 2H_2O$$
2 moles of NaOH : 1 mole of H_2SO_4

ACIDS AND CARBONATES

The reaction between carbonates and acids can be represented by:

$$CO_3^{2-} + 2H^+ \rightarrow H_2O + CO_2$$

This equation shows that one mole of CO_3^{2-} ions will always react with 2 moles of H^+ ions. Therefore, if a monobasic acid (e.g., HCl) is reacted with a carbonate, 2 moles of the acid will be required for every mole of the carbonate.

i.e., $$Na_2CO_3 + 2HCl \rightarrow H_2O + CO_2 + 2NaCl$$
2 moles of HCl : 1 mole of Na_2CO_3

Examples

(1) *1.400 g of pure anhydrous sodium carbonate were made up into 250 cm^3 of aqueous solution. 25.0 cm^3 of this solution required 24.50 cm^3 of a certain sample of hydrochloric acid. Calculate the molarity of the acid and its concentration in g dm^{-3}. If the remaining acid occupied 920 cm^3, how could it be made exactly decimolar?*

The equation for the reaction between hydrochloric acid and sodium carbonate is:

$$Na_2CO_3 + 2HCl \rightarrow 2NaCl + CO_2 + H_2O$$

i.e.,

2 moles of HCl react with 1 mole of Na_2CO_3
1 mole of Na_2CO_3 has a mass of 106 g.

Number of moles of Na_2CO_3 in 25 cm³ of
$$\text{solution} = \frac{1.400}{10} \times \frac{1}{106}$$
$$= 0.001321$$

Number of moles of HCl in 24.50 cm³ of
$$\text{solution} = 0.001321 \times 2$$
$$= 0.002642$$

Molarity is the number of moles in 1000 cm³ $= 0.002642 \times \dfrac{1000}{24.50}$
$$= 0.1077$$

1 mole of HCl has a mass of 36.5 g.

Concentration of the HCl solution $= 0.1077 \times 36.5$
$$= 3.93 \text{ g dm}^{-3}$$

Number of moles of HCl in 920 cm³ $= 0.1077 \times \dfrac{920}{1000}$
$$= 0.0991$$

When this solution is diluted to 0.1M it will still contain the same number of moles. But 1000 cm³ of decimolar solution contains 0.100 moles. Therefore $\dfrac{1000}{0.1000} \times 0.0991$ cm³ contains 0.0991 moles.

991 cm³ of 0.1M acid will contain the same number of moles as 920 cm³ of 0.1077M acid. Therefore 71 cm³ of distilled water must be added to the 920 cm³ of acid in order to convert it to exactly 0.1M acid.

(2) *3.000 g of a mixture of sodium carbonate and sodium chloride were made up to 250 cm³ of aqueous solution. 25.0 cm³ of this solution required 21.00 cm³ of 0.1050M hydrochloric acid (with methyl orange as indicator). Calculate the percentage by mass of sodium chloride in the mixture.*

In the titration sodium chloride is unchanged, while sodium carbonate reacts according to the equation:

$$Na_2CO_3 + 2HCl \rightarrow 2NaCl + H_2O + CO_2$$

i.e., 2 moles of HCl react with 1 mole of Na_2CO_3

or   1 mole of HCl reacts with 0.5 moles of $Na_2CO_3 = \dfrac{106}{2} = 53\,g$

$1000\,cm^3$ of 1M HCl reacts with 53 g of Na_2CO_3

$21.00\,cm^3$ of 0.1050M HCl reacts with $53 \times \dfrac{21.00}{1000} \times 0.1050\,g$

This is the mass of sodium carbonate in $25\,cm^3$, therefore the mass of sodium carbonate in $250\,cm^3$

$$= \dfrac{53 \times 21.00 \times 0.1050 \times 10}{1000}\,g$$

The percentage of sodium carbonate

$$= 53 \times \dfrac{21.00}{1000} \times 0.1050 \times 10 \times \dfrac{100}{3.00}$$

$$= 38.95$$

The percentage of sodium chloride $= 100 - 38.95 = 61.05$

(3) *1.340 g of a sample of ammonium chloride were boiled with excess of sodium hydroxide solution. The ammonia evolved was absorbed in $50.0\,cm^3$ of 0.5M sulphuric acid. The solution was then made up to $250\,cm^3$ with distilled water and $25\,cm^3$ of it required $25.10\,cm^3$ of 0.1M sodium hydroxide for neutralization. Calculate the percentage of ammonia (as NH_3) in the ammonium chloride.*

The equations involved are:

$$NH_4Cl + NaOH \rightarrow NaCl + H_2O + NH_3 \quad . \quad . \quad (i)$$
$$2NH_3 + H_2SO_4 \rightarrow (NH_4)_2SO_4 \quad . \quad . \quad . \quad (ii)$$
$$2NaOH + H_2SO_4 \rightarrow Na_2SO_4 + 2H_2O \quad . \quad . \quad (iii)$$

From equation *(iii)*, 1 mole of NaOH reacts with 0.5 moles of H_2SO_4. Therefore $25.10\,cm^3$ of 0.1M NaOH react with $2.51\,cm^3$ of 0.5M H_2SO_4.

The volume of excess 0.5M H_2SO_4 in $25\,cm^3$ of diluted solution $= 2.51\,cm^3$.

The total volume of excess 0.5M acid $= 25.10\,cm^3$.

The volume of acid reacting with the ammonia $= 50.0 - 25.10 = 24.90\,cm^3$.

From equation *(ii)*, 1 mole of H_2SO_4 reacts with 2 moles of NH_3.
Therefore $1000\,cm^3$ of 1M H_2SO_4 react with $2 \times 17 = 34\,g$ of NH_3.

$24.90\,cm^3$ of 0.5M H_2SO_4 react with $34 \times \dfrac{24.90}{1000} \times 0.5\,g$ of NH_3

The percentage of NH_3 in the ammonium chloride

$$= 34 \times \frac{24.90}{1000} \times 0.5 \times \frac{100}{1.340} = 31.6$$

(4) 1.600 g *of a metallic oxide of type MO were dissolved in* $100\,cm^3$ *1M hydrochloric acid. The resulting liquid was made up to* $500\,cm^3$ *with distilled water.* $25.0\,cm^3$ *of the solution then required* $21.02\,cm^3$ *of* $0.1020M$ *sodium hydroxide for neutralization. Calculate the mass of the oxide reacting with 1 mole of hydrochloric acid and hence the molar mass of the oxide and the relative atomic mass of the metal.*

The equations involved are:

$$MO + 2HCl \rightarrow MCl_2 + H_2O \qquad . \quad . \quad . \quad (i)$$
$$HCl + NaOH \rightarrow NaCl + H_2O \qquad . \quad . \quad . \quad (ii)$$

From equation (ii), 1 mole of HCl reacts with 1 mole of NaOH.

Number of moles of NaOH used $= 0.1020 \times \dfrac{21.02}{1000}$

Number of moles of HCl left in $25.0\,cm^3$ of the diluted solution is also

$$= 0.1020 \times \frac{21.02}{1000}$$

Total number of moles of excess acid $= 0.1020 \times \dfrac{21.02}{1000} \times 20$

$$= 0.04288$$

Number of moles of HCl in the $100\,cm^3$ of 1M acid $= 0.1000$

Number of moles of HCl reacting with the oxide $\begin{aligned}&= 0.100\\&- 0.04288\\&= 0.05712\end{aligned}$

i.e., 0.05712 moles of HCl have reacted with 1.600 g of the oxide, therefore

1 mole of HCl would react with $\dfrac{1.600}{0.05712}$ g of the oxide $= 28.01$ g

From equation (i), 0.5 moles of the oxide react with 1 mole of HCl. Therefore 1 mole of the oxide $= 2 \times 28.01 = 56.02$ g. This is the molar mass of the oxide and hence the relative atomic mass of $M = 56.02 - 16 = 40.02$.

(5) 5.032 g *of a dibasic organic acid, of anhydrous relative molecular mass 90, were made up to one* dm^3 *of aqueous solution.* $25.0\,cm^3$ *of this solution required* $19.97\,cm^3$ *of* $0.10M$ *sodium hydroxide for neutralization with phenolphthalein as indicator. Calculate the number of moles of water of crystallization per mole of the crystalline acid.*

When hydrated, the dibasic acid must be of the form $H_2A \cdot nH_2O$, reacting with sodium hydroxide according to the equation:

$$H_2A \cdot nH_2O + 2NaOH \rightarrow Na_2A + (n+2)H_2O$$

1 mole of acid reacts with 2 moles of sodium hydroxide, i.e., 1 mole of NaOH reacts with 0.5 mole of acid.

25 cm³ of the solution contained $\dfrac{5.032}{40}$ g of hydrated acid, therefore

19.97 cm³ of 0.10 M NaOH reacts with $\dfrac{5.032}{40}$ g of hydrated acid

1000 cm³ of 1 M NaOH reacts with $\dfrac{5.032}{40} \times \dfrac{1000}{19.97} \times 10$ g $= 62.99$ g

Therefore 63.0 g is half a mole of the hydrated acid and 126 g is 1 mole of the hydrated acid. This is made up of 1 mole of the anhydrous acid = 90 g and n moles of water = $18n$ g, i.e.,

$$90 + 18n = 126$$
$$n = 2$$

That is, there are 2 moles of water of crystallization per mole of the hydrated acid, $H_2A \cdot 2H_2O$.

(6) *Estimation of a mixture of sodium hydroxide and sodium carbonate by using two different indicators.*

The reaction between sodium hydroxide and hydrochloric acid occurs in a single stage, that between sodium carbonate and hydrochloric acid occurs in two stages, with the hydrogencarbonate ion as intermediate product.

$$NaOH + HCl \rightarrow NaCl + H_2O \quad \quad \quad (i)$$
$$Na_2CO_3 + HCl \rightarrow NaCl + NaHCO_3 \quad \quad (ii)$$
$$NaHCO_3 + HCl \rightarrow NaCl + H_2O + CO_2 \quad (iii)$$

or ionically
$$OH^- + H^+ \rightarrow H_2O$$
$$CO_3^{2-} + H^+ \rightarrow HCO_3^-$$
$$HCO_3^- + H^+ \rightarrow H_2O + CO_2$$

With phenolphthalein as indicator, the end point (pink → colourless) is registered at the completion of reactions (*i*) and (*ii*). With methyl orange as indicator, the end point (yellow → pink) is registered at the completion of all three reactions. Since the amount of acid required for reactions (*ii*) and (*iii*) is the same, and reaction (*iii*) is the difference between the titrations with methyl orange and phenolphthalein, it follows that twice this difference is the amount of acid which titrates the carbonate. The rest titrates the sodium hydroxide.

Example

25.0 cm³ of a solution containing sodium hydroxide and sodium carbonate required 21.05 cm³ of 1M HCl with phenolphthalein as indicator and 27.85 cm³ of the same acid with methyl orange as indicator. Calculate the concentration of each compound in g dm⁻³ of the solution.

From the equations the acid required to titrate the sodium carbonate in 25.0 cm³ is 2(27.85 − 21.50) cm³, or 12.70 cm³. Consequently, the acid required to titrate the sodium hydroxide in the same volume of solution is (27.85 − 12.70) cm³, or 15.15 cm³.

From the equations, 1 mole of HCl reacts with 0.5 mole of Na_2CO_3 = 53 g, i.e.,

1000 cm³ of 1M HCl reacts with 53 g of Na_2CO_3

12.70 cm³ reacts with $53 \times \dfrac{12.70}{1000}$ g

This is the mass of sodium carbonate in 25.0 cm³ of solution, therefore the mass in 1 dm³

$$= 53 \times \dfrac{12.70}{1000} \times 40 = 27.6 \, g$$

From the equation 1 mole of HCl reacts with 1 mole of NaOH = 40 g, i.e.,

1000 cm³ of 1M HCl reacts with 40 g of NaOH

15.15 cm³ reacts with $40 \times \dfrac{15.15}{1000}$ g

This is the mass of sodium hydroxide in 25.0 cm³ of solution, therefore the mass in 1 dm³

$$= 40 \times \dfrac{15.15}{1000} \times 40 = 24.2 \, g$$

The same method can be applied to the estimation of sodium carbonate and hydrogencarbonate together. With phenolphthalein as indicator, the end point (pink → colourless) is given when all the carbonate is converted to hydrogencarbonate.

$$CO_3^{2-} + H^+ \rightarrow HCO_3^- \qquad . \qquad . \qquad . \qquad (i)$$

With methyl orange the end point appears (yellow → pink) when the original hydrogencarbonate and that produced in (i) are both converted to carbon dioxide and water.

From (i), $\qquad HCO_3^- + H^+ \rightarrow H_2O + CO_2 \qquad . \qquad . \qquad (ii)$
Original $\qquad HCO_3^- + H^+ \rightarrow H_2O + CO_2 \qquad . \qquad . \qquad (iii)$

Since the amount of acid required for reactions (i) and (ii) is the same, twice the phenolphthalein titration is the amount of acid which titrates the carbonate, the rest titrates the original hydrogencarbonate. The rest of the calculation is similar to the above, from the mole ratios:

1 mole Na_2CO_3 : 2 moles HCl;   1 mole $NaHCO_3$: 1 mole HCl

(7) 1.600 g of an acid, of relative molecular mass 118, were made up to 250 cm^3 of aqueous solution. 25.0 cm^3 of this solution required 27.10 cm^3 of 0.10M sodium hydroxide for neutralization. Calculate the mass of the acid reacting with 1 mole of sodium hydroxide and hence the basicity of the acid.

25.0 cm^3 of the solution contains 0.160 g of the acid, therefore

27.10 cm^3 of 0.10M NaOH react with 0.160 g of the acid

1000 cm^3 of 1M NaOH react with $0.160 \times \dfrac{1000}{27.10} \times 10 = 59.03$ g

But this is half a mole of acid, therefore 1 mole of the acid reacts with 2 moles of the sodium hydroxide and the acid is dibasic.

Problems on Acids and Alkalis

(Relative atomic masses will be found on page 132)

STANDARDIZATIONS

(1) 100 cm^3 of concentrated hydrochloric acid were diluted to 1 dm^3 with distilled water. 26.8 cm^3 of this diluted acid were needed to neutralize 25 cm^3 of 0.5M sodium carbonate solution, with methyl orange as indicator. What is the concentration in g dm^{-3} of the original acid?

(2) To 25 cm^3 of a solution of sodium hydroxide, 50 cm^3 of a 0.5M hydrochloric acid solution were added and the resulting solution required 22.3 cm^3 of 0.25M NaOH for neutralization. Calculate the original concentration of the sodium hydroxide solution in g dm^{-3}.

(3) In standardizing a roughly 0.05M H_2SO_4 solution, 1.325 g of pure anhydrous sodium carbonate was made up to 250 cm^3 and 25 cm^3 of the solution needed 23.5 cm^3 of the acid. Calculate the concentration of the acid in g dm^{-3}. If 906 cm^3 of the acid were left after the titration, how many cm^3 of water must be added to the remaining acid to make it exactly 0.05M?

(4) Calculate the molarity of a sodium hydroxide solution from the following data: 2.01 g of $KH_3(C_2O_4)_2 \cdot 2H_2O$ needed 26.2 cm^3 of the sodium hydroxide solution for neutralization, phenolphthalein being used as indicator.

(5) Hydrochloric acid was standardized in the following way: 1.010 g of pure calcium carbonate was allowed to react with 50 cm^3

of the hydrochloric acid. The excess of the acid was neutralized by 24.6 cm^3 of a sodium hydroxide solution. 25 cm^3 of this sodium hydroxide solution needed 23.7 cm^3 of the hydrochloric acid solution for neutralization. Calculate the molarity of the acid.

MIXTURES EXCLUDING NH_4^+

(6) 100 cm^3 of vinegar at 15°C were diluted to 250 cm^3 with distilled water. 25 cm^3 of the diluted solution required 16.9 cm^3 of 0.5M NaOH for neutralization with phenolphthalein as indicator. Assuming that all the acidity of vinegar is caused by ethanoic acid, calculate the percentage by mass of this acid in the original vinegar. (Density of vinegar at 15°C is 1.02 g cm^{-3}.)

(7) 20 cm^3 of a solution containing sodium hydroxide and sodium carbonate required 19.2 cm^3 of 0.5M HCl with phenolphthalein as indicator. With methyl orange, a further 5.1 cm^3 of the acid were needed. What is the concentration of each compound in the original solution, in grams of anhydrous solid per dm^3?

(8) 2.50 g of a mixture of anhydrous sodium carbonate and sodium chloride were made up to 250 cm^3 with distilled water. 25 cm^3 of this solution required 20.0 cm^3 of 0.1M HCl, with methyl orange as indicator. Calculate the percentage by mass of sodium chloride in the mixture.

(9) Anhydrous sodium carbonate contaminated with sodium hydrogencarbonate was made up to 250 cm^3 of solution. 25 cm^3 of this solution required 11.2 cm^3 of 1M HCl with phenolphthalein as indicator and 24.5 cm^3 of the same acid with methyl orange as indicator. Calculate the percentage by mass of sodium hydrogencarbonate in the mixture.

(10) 25 cm^3 of solution of sodium hydroxide and potassium hydroxide containing 5 g of solid per dm^3 required 24.2 cm^3 of 0.1M HCl for neutralization. Calculate the concentration of each compound in g dm^{-3} of solution.

(11) 25 cm^3 of a solution containing sodium hydroxide and sodium carbonate required 28.0 cm^3 of decimolar hydrochloric acid for neutralization with methyl orange indicator. The carbonate ions were removed from 25 cm^3 of the solution by the addition of excess of barium chloride solution. The resulting mixture was then titrated slowly with the same acid, and with phenolphthalein as indicator. 18.0 cm^3 of the acid were needed. Calculate the mass of sodium hydroxide and sodium carbonate (anhydrous) per dm^3 of the solution.

(12) 1.500 g of a sample of limestone were dissolved in 50 cm^3 of 1M hydrochloric acid. The resulting solution was made up to 250 cm^3 with distilled water. 25 cm^3 of this solution required 21.05 cm^3 of 0.1M sodium hydroxide for neutralization. Assuming all the basic material

in the rock to be calcium carbonate, calculate the percentage of calcium carbonate present.

(13) 7.12 g of an indigestion mixture were added to 50 cm^3 of 1M hydrochloric acid and when the reaction was complete, the solution was diluted to 250 cm^3 with distilled water. 25 cm^3 of this diluted solution required 24.4 cm^3 of 0.1M sodium hydroxide for neutralization of the excess acid. Assuming all the basic material in the indigestion mixture to be magnesium hydroxide, calculating the percentage of magnesium hydroxide in the mixture.

MIXTURES INVOLVING NH_4^+

(14) 5.000 g of ammonium chloride contaminated with sodium chloride were boiled with 100 cm^3 of 2M NaOH until no more ammonia was evolved. The residual solution was made up to 250 cm^3 with water and 25 cm^3 of this required 22.4 cm^3 of 0.5M HCl for neutralization. What was the mass of sodium chloride in the ammonium chloride?

(15) 25 cm^3 of a solution containing 5.00 g of ammonium chloride per dm^3 were boiled with excess sodium hydroxide solution. The ammonia was absorbed by passing it into 25 cm^3 of 0.1M sulphuric acid. What volume of 0.1M sodium hydroxide solution would be needed to titrate the excess sulphuric acid?

(16) Calculate the percentage of ammonium sulphate in a sample of this compound contaminated with sodium sulphate from the following data: 1.65 g of the ammonium sulphate was made up to 250 cm^3 of aqueous solution. 25 cm^3 of this were boiled with 50 cm^3 0.1M NaOH and the excess alkali neutralized by 25.4 cm^3 of 0.1M HCl.

(17) Calculate the percentage by mass of each salt in a mixture of ammonium chloride and ammonium sulphate from the following data: 6.000 g of the mixture were made up to 1 dm^3 of aqueous solution. 25 cm^3 of this solution were boiled with 50 cm^3 of 0.1M NaOH until ammonia was no longer evolved. The excess alkali required 25.6 cm^3 of 0.1M HCl to neutralize it.

(18) 5.000 g of a sample of ammonium chloride were boiled with excess sodium hydroxide solution and the evolved ammonia was absorbed in 120 cm^3 of 0.5M H_2SO_4. The excess acid was made up to 250 cm^3 with water and 25 cm^3 of the solution then required 26.4 cm^3 of 0.1M NaOH for neutralization. Calculate the percentage of ammonia, as NH_3, in the sample.

WATER OF CRYSTALLIZATION

(19) 8.58 g of washing soda were made up to 250 cm^3 of aqueous solution. 25 cm^3 of this solution required 30.0 cm^3 of 0.2M HCl for neutralization with methyl orange as indicator. Calculate x in the formula, $Na_2CO_3 \cdot xH_2O$, for washing soda.

(20) 1.260 g of a dibasic organic acid, of anhydrous relative molecular mass 90, were made up to 250 cm^3 with water. 25 cm^3 of this solution required 19.95 cm^3 0.1M NaOH for neutralization. Calculate the number of moles of water of crystallization per mole of the crystalline acid.

(21) A potassium salt has the formula
$$H_2C_2O_4 \cdot KHC_2O_4 \cdot xH_2O$$
Calculate x from the following data: 1.923 g of the potassium salt was neutralized by 22.7 cm^3 of 1M NaOH, phenolphthalein as indicator.

(22) 10 g of copper(II) sulphate crystals were dissolved in water and boiled with 120 cm^3 of 1M NaOH. The precipitated oxide was filtered off and filtrate and washings made up to 250 cm^3. 20 cm^3 of this solution required 32.0 cm^3 of 0.05M H_2SO_4. Calculate the number of moles of water of crystallization in one mole of the crystals.

MOLES

(23) 0.800 g of a metallic oxide of the type MO was dissolved in 50 cm^3 1M HCl. The resulting liquid was made up to 250 cm^3 with water and 25 cm^3 of this solution required 21.4 cm^3 of 0.1M NaOH for neutralization. Calculate the mass of the oxide reacting with 1 mole of HCl and hence the relative molecular mass of the oxide and the relative atomic mass of M.

(24) Metal Z has a relative atomic mass of 24. From the data calculate the number of moles of Z combining with 1 mole of H_2SO_4, write a balanced equation for the reaction and deduce the charge on the metal ion in the resulting solution. 0.575 g of Z was allowed to react with 50 cm^3 of 0.5M H_2SO_4. The excess acid required 29.9 cm^3 of 0.096M KOH for neutralization.

(25) From the following data calculate the mass of a dibasic organic acid which reacts with 1 mole of sodium hydroxide and hence deduce the relative molecular mass of the acid. 7.30 g of the acid were made up to 250 cm^3 of aqueous solution. 25 cm^3 of this solution required 22.5 cm^3 of 0.5M sodium hydroxide for neutralization.

(26) Calculate the relative molecular mass of the carbonate of a divalent metal X and hence the relative atomic mass of X from the data: 1.000 g of the anhydrous normal carbonate of X was added to 50 cm^3 of 1M HCl. The excess acid required 30.0 cm^3 of 1M NaOH for neutralization.

(27) 1.00 g of an anhydrous organic acid of molar mass 90 g was made up to 250 cm^3 of aqueous solution and 25 cm^3 of this solution needed 22.2 cm^3 of 0.1M KOH for neutralization. What mass of the acid reacts with 1 mole of KOH and hence what is the basicity of the acid?

2

Redox Titrations

Theory

Oxidation and reduction always occur simultaneously in the same reaction. Such reactions are known as redox reactions.

Oxidation is any change involving loss of electrons.

Reduction is any change involving gain of electrons.

It follows that any chemical species losing electrons, and becoming oxidized, must give these electrons to another species which is therefore reduced.

Any species donating electrons is a reducing agent.

Any species accepting electrons is an oxidizing agent.

The three main types of redox titrations are based on the use of the following reagents:

   (i) potassium manganate(VII),
   (ii) potassium dichromate,
   (iii) sodium thiosulphate.

POTASSIUM MANGANATE(VII)

Theory

Potassium manganate(VII) in acidic conditions (produced by the presence of dilute sulphuric acid) acts as an oxidizing agent according to the ionic half equation:

$$MnO_4^- + 8H^+ + 5e^- \rightarrow Mn^{2+} + 4H_2O$$

Thus each MnO_4^- ion can accept five electrons.

Acidic potassium manganate(VII) solution is used chiefly in redox titrations with iron(II) salts, ethanedioates (oxalates) and hydrogen peroxide solution.

The ionic half equation for iron(II) ions donating electrons is:

$$Fe^{2+} \rightarrow Fe^{3+} + e^-$$

Therefore 1 mole of MnO_4^- ions (or $KMnO_4$) will oxidize 5 moles of Fe^{2+} ions. The $5Fe^{2+}$ can come from a soluble iron(II) salt, e.g., $5FeSO_4$ or from dissolving 5 moles of iron metal in dilute sulphuric acid or from the reduction of 5 moles of iron(III) ions.

The ionic half equation for ethanedioate ions is:

$$C_2O_4^{2-} \rightarrow 2CO_2 + 2e^-$$

Therefore 2 moles of MnO_4^- will oxidize 5 moles of $C_2O_4^{2-}$ ions, e.g., $5H_2C_2O_4$ or $5Na_2C_2O_4$.

The ionic half equation for hydrogen peroxide acting as a reducing agent is:

$$H_2O_2 \rightarrow 2H^+ + O_2 + 2e^-$$

Therefore 2 moles of MnO_4^- will oxidize 5 moles of H_2O_2.

It is these mole ratios, derived from the ionic half equations, which form the basis of the calculations involving potassium manganate(VII) as a titrating agent.

Examples

(1) *9.85 g of pure ammonium iron(II) sulphate were made up to 250 cm³ of solution in cold, boiled-out, dilute sulphuric acid. 25.0 cm³ of the solution reacted completely with 24.75 cm³ of a potassium manganate(VII) solution. Calculate the molarity of the solution, its concentration in $g\, dm^{-3}$ and the dilution needed to convert the remaining 1800 cm³ of it to exactly 0.02M concentration.*

The ionic half equations for the oxidation of the iron(II) salt by the manganate(VII) are:

$$MnO_4^- + 8H^+ + 5e^- \rightarrow Mn^{2+} + 4H_2O$$
$$Fe^{2+} \rightarrow Fe^{3+} + e^-$$

1 mole of ammonium iron(II) sulphate, $FeSO_4.(NH_4)_2SO_4.6H_2O$, is 392 g.

25.0 cm³ of the iron(II) salt solution contain $\dfrac{9.85}{10} \times \dfrac{1}{392}$ moles of Fe^{2+}, as $1MnO_4^-$ reacts with $5Fe^{2+}$, 24.75 cm³ of the manganate(VII) solution must contain $\dfrac{9.85}{10} \times \dfrac{1}{392} \times \dfrac{1}{5}$ moles of MnO_4^-.

1 dm³ of the manganate(VII) solution contains

$$\frac{9.85}{10} \times \frac{1}{392} \times \frac{1}{5} \times \frac{1000}{24.75} = 0.0203 \text{ moles}$$

The potassium manganate(VII) solution is 0.0203 M
1 mole of potassium manganate(VII) is 158 g. Therefore the concentration of the potassium manganate(VII) solution = 0.0203 × 158 = 3.21 g dm^{-3}

1800 cm³ of this solution contains

$$0.0203 \times \frac{1800}{1000} = 0.03654 \text{ moles of } KMnO_4$$

When this solution is diluted to make a 0.02 M solution, it will still contain 0.03654 moles.
The volume of a 0.02 M $KMnO_4$ solution containing 0.03654 moles

$$= 1000 \times \frac{0.03654}{0.02} = 1827 \text{ cm}^3$$

i.e., 27 cm³ of distilled water must be added to the 1800 cm³ of potassium manganate(VII) solution in order to make it exactly 0.02 M.

(2) *25.00 cm³ of a solution containing ethanedioic acid and sodium ethanedioate required 14.75 cm³ of 0.1 M sodium hydroxide solution for neutralization, and 30.5 cm³ of 0.0205 M potassium manganate(VII) solution for oxidation in acidic conditions at about 70°C. Calculate the number of grams of each (anhydrous) constituent per dm³ of the solution.*

The sodium hydroxide reacts with the free ethanedioic acid only. The reaction is:

$$H_2C_2O_4 + 2NaOH \rightarrow Na_2C_2O_4 + 2H_2O$$

1 mole of NaOH reacts with 0.5 moles of $H_2C_2O_4$ = 45 g
1000 cm³ of 1 M NaOH reacts with 45 g of $H_2C_2O_4$

$$14.75 \text{ cm}^3 \text{ of } 0.1 \text{ M NaOH reacts with } 45 \times \frac{14.75}{1000} \times 0.1 \text{ g}$$

This is the mass of ethanedioic acid in 25.00 cm³ of solution, therefore the mass of ethanedioic acid in 1 dm³

$$= 45 \times \frac{14.75}{1000} \times 0.1 \times 40$$

$$= 2.66 \text{ g}$$

The manganate(VII) oxidizes all the ethanedioate present according to the ionic half equations:

$$MnO_4^- + 8H^+ + 5e^- \rightarrow Mn^{2+} + 4H_2O$$
$$C_2O_4^{2-} \rightarrow 2CO_2 + 2e^-$$

1 mole of MnO_4^- oxidizes 2.5 moles of ethanedioate ions
1000 cm³ of 1M $KMnO_4$ oxidizes 2.5 moles of ethanedioate ions
30.15 cm³ of 0.0205M $KMnO_4$ oxidizes $2.5 \times \dfrac{30.15}{1000} \times 0.0205$
moles, i.e.,
the total ethanedioate ion present in 25.00 cm³ = 0.001545 moles.
But the number of moles of ethanedioate from ethanedioic acid in 25.00 cm³

$$= \dfrac{2.66}{40} \times \dfrac{1}{90} = 0.0007375$$

Therefore the number of moles of ethanedioate ions from sodium ethanedioate

$$= 0.001545 - 0.0007375 = 0.0008075$$

But 1 mole of $Na_2C_2O_4$ is 134 g, therefore
the mass of $Na_2C_2O_4$ in 25.00 cm³ $= 0.0008075 \times 134$ g
and
the mass in 1000 cm³ solution $= 0.0008075 \times 134 \times 40$
$= 4.33$ g

(3) *12.5 cm³ of a given solution of hydrogen peroxide were diluted to 500 cm³ with distilled water. 25.0 cm³ of this diluted solution then required 22.50 cm³ of 0.02M potassium manganate(VII) solution for titration in acidic conditions. Calculate the concentration of the hydrogen peroxide solution in g dm^{-3} and express its rating as 'volume' solution, assuming s.t.p. conditions.*

The ionic half equations are:

$$MnO_4^- + 8H^+ + 5e^- \rightarrow Mn^{2+} + 4H_2O$$
$$H_2O_2 \rightarrow 2H^+ + O_2 + 2e^-$$

2 moles of MnO_4^- will oxidize 5 moles of H_2O_2
1 mole of MnO_4^- will oxidize 2.5 moles of $H_2O_2 = 2.5 \times 34$ g
1000 cm³ of 1M $KMnO_4$ will oxidize 2.5×34 g of H_2O_2
22.50 cm³ of 0.02M $KMnO_4$ will oxidize $2.5 \times 34 \times \dfrac{22.50}{1000} \times 0.02$ g

This is the mass of hydrogen peroxide in 25 cm³ of the diluted solution, therefore the mass in 1000 cm³

$$= 2.5 \times 34 \times \dfrac{22.50}{1000} \times 0.02 \times 40 = 1.53 \text{ g}$$

Therefore, the original hydrogen peroxide solution contained

$$1.53 \times \frac{500}{12.5} = 61.2 \text{ g dm}^{-3}$$

To express this as a 'volume' solution, we require the following reaction for the heating of hydrogen peroxide:

$$2H_2O_2 \rightarrow 2H_2O + O_2$$
$$68 \text{ g} \qquad\qquad 22.4 \text{ dm}^3 \text{ at s.t.p.}$$

From this, 61.2 g of hydrogen peroxide produce $22.4 \times \frac{61.2}{68}$ or 20.2 dm³ of oxygen, referred to s.t.p., when heated. That is, it produces 20.2 times its own volume of oxygen (referred to s.t.p.) when heated.

(4) 25.0 cm³ *of a solution containing iron(II) and iron(III) sulphates required 18.50 cm³ of 0.02M potassium manganate(VII) solution for oxidation in acidic conditions. After complete reduction by zinc amalgam, 25.0 cm³ of the solution required 32.60 cm³ of the same manganate(VII) solution. Calculate the number of grams of each anhydrous sulphate per dm³ of the solution.*

The oxidation of the iron(II) ions by the manganate(VII) ions is represented by:

$$MnO_4^- + 8H^+ + 5e^- \rightarrow Mn^{2+} + 4H_2O$$
$$Fe^{2+} \rightarrow Fe^{3+} + e^-$$

1 mole of MnO_4^- will oxidize 5 moles of Fe^{2+}, i.e.,

1 mole of $KMnO_4$ will oxidize 5 moles of $FeSO_4 = 5 \times 152$ g and 5 moles of $FeSO_4$ could be formed from the reduction of 2.5 moles of $Fe_2(SO_4)_3 = 2.5 \times 400$ g.

From the first titration,

1000 cm³ of 1 M $KMnO_4$ will oxidize 5×152 g of $FeSO_4$, therefore the mass of iron(II) sulphate in 25.0 cm³ of solution

$$= 5 \times 152 \times \frac{18.50}{1000} \times 0.02 \text{ g}$$

mass of iron(II) sulphate in 1 dm³

$$= 5 \times 152 \times \frac{18.50}{1000} \times 0.02 \times 40 = 11.25 \text{ g}$$

After complete reduction, the iron(II) ions derived from the original iron(III) ions are also oxidized by the manganate(VII) solution, therefore the iron(III) salt produces sufficient iron(II) ions to react with (32.60 − 18.50) = 14.10 cm³ of 0.02M $KMnO_4$ solution.

1 mole of $KMnO_4$ is equivalent to 2.5×400 g of iron(III) sulphate

100 cm^3 of 1M $KMnO_4$ is equivalent to 2.5×400 g, therefore the mass of iron(III) sulphate in 25.0 cm^3 of solution

$$= 2.5 \times 400 \times \frac{14.10}{1000} \times 0.02 \text{ g}$$

Mass of iron(III) sulphate in 1 dm^3

$$= 2.5 \times 400 \times \frac{14.10}{1000} \times 0.02 \times 40 = 11.28 \text{ g}$$

Problems on Potassium Manganate(VII)
(*Relative atomic masses will be found on page 132*)

(1) A roughly 0.02M potassium manganate(VII) solution was standarized against exactly 0.1M ammonium iron(II) sulphate solution. 25 cm^3 of the solution of the iron(II) salt were oxidized by 24.6 cm^3 of the manganate(VII) solution. What is the molarity of the manganate(VII) solution?

(2) A standardization of potassium manganate(VII) solution yielded the following data: 0.160 g of the potassium salt,

$$KHC_2O_4 . H_2C_2O_4 . 2H_2O$$

needed 24.5 cm^3 of the manganate(VII) solution. What is the molarity of the manganate(VII) solution?

(3) 1.600 g of ethanedioic acid crystals, $H_2C_2O_4 . 2H_2O$, was made up to 250 cm^3 of aqueous solution and 25 cm^3 of this solution needed 26.2 cm^3 of a manganate(VII) solution for oxidation. Calculate the molarity of the manganate(VII) solution and its concentration in g dm^{-3}.

(4) Calculate the concentration in g dm^{-3} of a sodium ethanedioate solution, 25 cm^3 of which were oxidized in acid solution by 28.5 cm^3 of a potassium manganate(VII) solution containing 2.500 g dm^{-3}.

(5) Calculate the percentage of iron in a sample of iron wire from the following data: 1.400 g of the wire was dissolved in excess of dilute sulphuric acid and the solution made up to 250 cm^3. 25 cm^3 of this solution needed 25.37 cm^3 of 0.0196M $KMnO_4$ for oxidation.

(6) Calculate x in the formula $FeSO_4 . xH_2O$ from the following data: 24.400 g of iron(II) sulphate crystals were made up to 1 dm^3 of aqueous solution acidified with sulphuric acid. 25 cm^3 of this solution required 20.0 cm^3 of 0.022M $KMnO_4$.

(7) Calculate the percentage by mass of iron(III) in a salt from the following data: 25.000 g of the salt were dissolved in water and reduced to an iron(II) solution by zinc and dilute sulphuric acid. The mixture was filtered and the filtrate and washings made up to 1 dm^3. 20 cm^3 of this solution required 20.7 cm^3 of 0.01M KMnO$_4$ for oxidation.

(8) Calculate the masses of iron(II) and iron(III) per dm^3 of a solution from the following data: 25 cm^3 of the solution required 18.6 cm^3 of 0.02M KMnO$_4$ for oxidation. After reduction by zinc and acid, 25 cm^3 of the solution required 34.6 cm^3 of the same solution of KMnO$_4$.

(9) Calculate the masses of anhydrous iron(II) and iron(III) sulphate present in 1 dm^3 of solution from the following results: 25 cm^3 of the solution were oxidized by 23.0 cm^3 of 0.02M KMnO$_4$. After reduction by zinc and dilute sulphuric acid, 25 cm^3 of the solution required 40.0 cm^3 of the same manganate(VII) solution.

(10) Calculate the masses of anhydrous ethanedioic acid and ammonium ethanedioate per dm^3 of a given solution from the data: 25 cm^3 of the solution required 17.6 cm^3 of 0.1M NaOH for neutralization, phenolphthalein as indicator. 25 cm^3 of the solution required 38.7 cm^3 of 0.0196M KMnO$_4$ for oxidation when acidified.

(11) 1.500 g of a mixture of anhydrous sodium ethanedioate and ethanedioic acid crystals was made up to 100 cm^3 of aqueous solution. 25 cm^3 of this solution required 19.8 cm^3 of 0.1M NaOH for neutralization, phenolphthalein as indicator. How many cm^3 of 0.01M KMnO$_4$ will be necessary to oxidize 25 cm^3 of the original solution in the presence of excess dilute sulphuric acid?

(12) Calculate the percentage of calcium in calcspar from the following data: 1.25 g of calcspar was dissolved in dilute hydrochloric acid and calcium ethanedioate was precipitated from the neutralized solution by ammonium ethanedioate. The washed precipitate was dissolved by addition of hot dilute sulphuric acid and made up to 250 cm^3. 25 cm^3 of this solution needed 24.0 cm^3 of 0.0208M KMnO$_4$ for oxidation.

(13) 52.3 cm^3 of sodium nitrite solution, added from a burette were needed to discharge the colour of 25 cm^3 of an acidified 0.02M KMnO$_4$ solution. What was the concentration of the nitrite solution in grams of anhydrous salt per dm^3? Why was the nitrite in the burette?

(14) 100 cm^3 of solution of hydrogen peroxide were diluted to 1 dm^3 with water. 25 cm^3 of this solution, when acidified with dilute sulphuric acid, reacted with 47.8 cm^3 of 0.02M KMnO$_4$. What is the concentration of the original hydrogen peroxide solution in g dm^{-3}? What is the 'volume' rating of this solution (referred to s.t.p.)?

(15) A solution of ethanedioic acid was made up and it was calculated that 22.5 cm³ of certain potassium manganate(VII) solution would oxidize 25 cm³ of the ethanedioic acid solution. In practice it was found that only 16.1 cm³ of the manganate(VII) solution were needed. The error lay in the fact that the calculation was based on the formula of the anhydrous acid whereas the crystalline acid was actually weighed out. What is the value of x in the formula $H_2C_2O_4 \cdot xH_2O$ for the crystalline acid?

(16) Calculate the percentage of manganese(IV) oxide in a sample of pyrolusite from the results: 25 cm³ of a certain ethanedioic acid solution required 47.6 cm³ of 0.02M $KMnO_4$ for oxidation. To 25 cm³ of this ethanedioic acid solution, 0.140 g of finely powdered pyrolusite was added and the mixture was boiled with dilute sulphuric acid till no black particles remained. The solution then required 23.2 cm³ of 0.02M $KMnO_4$ for oxidation.

$$MnO_2 + C_2O_4^{2-} + 4H^+ \rightarrow Mn^{2+} + 2H_2O + 2CO_2$$

(17) Calculate the concentration in g dm⁻³ of a solution of potassium chlorate(V) from the following data: 50.0 cm³ of a solution of ammonium iron(II) sulphate (acidified) were boiled for ten minutes with 25.0 cm³ of the potassium chlorate solution. After cooling, the excess iron(II) salt was oxidized by 24.6 cm³ of 0.02M $KMnO_4$ solution. 25.0 cm³ of the same acidified ammonium iron(II) sulphate solution required 24.2 cm³ of 0.02M $KMnO_4$.

$$6Fe^{2+} + 6H^+ + ClO_3^- \rightarrow Cl^- + 3H_2O + 6Fe^{3+}$$

(18) 1.27 g of iron(II) ethanedioate were made up to 250 cm³ of acidified aqueous solution. 25 cm³ of this solution reacted completely with 26.5 cm³ of 0.02M potassium manganate(VII) solution. Calculate the mole ratio of $KMnO_4$ to FeC_2O_4 taking part in this reaction and suggest an equation for the reaction between $KMnO_4$ and FeC_2O_4 in the presence of H_2SO_4. Rewrite this equation in its simplest ionic terms.

(19) Hydroxyammonium ions, in acid solution, reduce iron(III) ions to iron(II) ions. Determine the mole ratio of hydroxyammonium ions (NH_3OH^+) to iron(III) ions (Fe^{3+}) taking part in the reaction from the following data: 1.00 g of hydroxyammonium sulphate were made up to 250 cm³ of aqueous solution. 25 cm³ of this solution were added to excess ammonium iron(III) sulphate and the iron(II) ions produced required 24.4 cm³ of 0.02M potassium manganate(VII) solution.

POTASSIUM DICHROMATE

Theory

An acidified solution of potassium dichromate acts as a powerful oxidizing agent according to the ionic half equation:

$$Cr_2O_7^{2-} + 14H^+ + 6e^- \rightarrow 2Cr^{3+} + 7H_2O$$

The dichromate is chiefly used in the oxidation of iron(II) salts,

$$Fe^{2+} \rightarrow Fe^{3+} + e^-$$

It has one particular advantage over potassium manganate(VII) in that it can be used to estimate iron(II) ions in the presence of chloride ions. Potassium manganate(VII) will oxidize both the iron(II) ions and the chloride ions,

$$2Cl^- \rightarrow Cl_2 + 2e^-$$

but dichromate ions will not react with chloride ions.

From the above ionic half equations it can be seen that:

1 mole of $K_2Cr_2O_7$ will oxidize 6 moles of Fe^{2+} ions.

Problems on Potassium Dichromate

(Relative atomic masses will be found on page 132)

(20) 12.0 g of ammonium iron(II) sulphate crystals were made up to 250 cm³ of acidified aqueous solution. 25 cm³ of this solution required 25.5 cm³ of 0.02M potassium dichromate for oxidation. Calculate x in the formula $FeSO_4 \cdot (NH_4)_2SO_4 \cdot xH_2O$

(21) 3.00 g of a sample of haematite (Fe_2O_3) were dissolved in concentrated hydrochloric acid and the solution diluted to 250 cm³. 25 cm³ of this solution, after reduction with tin(II) chloride, required 26.6 cm³ of 0.02M potassium dichromate for oxidation. Calculate the percentage of iron(III) oxide in the ore.

SODIUM THIOSULPHATE

Theory

Sodium thiosulphate, $Na_2S_2O_3$, reacts as a reducing agent towards iodine. Like all reducing agents, sodium thiosulphate acts as an electron donor. Expressed in ionic terms, the essential change is:

$$2S_2O_3^{2-} \rightarrow S_4O_6^{2-} + 2e^-$$

Simultaneously, the iodine (in potassium iodide solution) is reduced by accepting the electrons: $I_2 + 2e^- \to 2I^-$. The combined effect is:

$$2S_2O_3^{2-} + I_2 \to S_4O_6^{2-} + 2I^-$$

Consequently, 1 mole of $S_2O_3^{2-}$ ions will reduce 0.5 moles of I_2, or 1000 cm^3 of 1M $Na_2S_2O_3$ (containing 248 g of $Na_2S_2O_3.5H_2O$ crystals) will reduce 0.5 moles of iodine (127 g). (Iodine is sparingly soluble in water and is therefore dissolved in a solution of potassium iodide in which it forms an unstable ion, $I_2 + I^- \rightleftharpoons I_3^-$.)

Standard sodium thiosulphate solution is used to estimate iodine or, more frequently, any oxidizing agent which liberates iodine quantitatively from potassium iodide solution. The ionic half equation for this is:

$$2I^- \to I_2 + 2e^-$$

Examples

(1) *In a standardization of sodium thiosulphate solution, 25.0 cm^3 of 0.0204M potassium manganate(VII) solution were added to excess of acidified potassium iodide solution. The iodine liberated required 24.40 cm^3 of sodium thiosulphate solution for reduction. Calculate the molarity of the sodium thiosulphate solution. If 1750 cm^3 of it remain, how may it be made exactly decimolar?*

The liberated iodine acts merely as a connecting link between the manganate(VII) and the sodium thiosulphate solutions. It is not necessary to find the quantity of iodine produced.

The ionic half equations for the liberation of iodine are:

$$MnO_4^- + 8H^+ + 5e^- \to Mn^{2+} + 4H_2O$$
$$2I^- \to I_2 + 2e^-$$

The iodine then reacts with the thiosulphate according to the equation:

$$2S_2O_3^{2-} + I_2 \to S_4O_6^{2-} + 2I^-$$

i.e., 1 mole of MnO_4^- liberates 2.5 moles of I_2 which will then react with 5 moles of $S_2O_3^{2-}$. Therefore 1 mole of MnO_4^- can be said to be equivalent to 5 moles of $S_2O_3^{2-}$.

Number of moles of MnO_4^- in 25.0 cm^3 of solution $= 0.0204 \times \dfrac{25.0}{1000}$

Number of moles of $S_2O_3^{2-}$ in 24.40 cm^3 $= 0.0204 \times \dfrac{25.0}{1000} \times 5$

Number of moles of $S_2O_3^{2-}$ in 1000 cm^3

$$= 0.0204 \times \frac{25.0}{1000} \times 5 \times \frac{1000}{24.40} = 0.1045$$

The sodium thiosulphate solution is 0.1045 M.

1750 cm^3 of this solution will contain $0.1045 \times \frac{1750}{1000}$ moles. When the solution has been diluted to make it decimolar it will still contain the same number of moles. The volume of 0.1 M $Na_2S_2O_3$ containing this number of moles

$$= \frac{1000}{0.1} \times 0.1045 \times \frac{1750}{1000} = 1829 \text{ cm}^3$$

Therefore, 79 cm^3 of distilled water must be added to the 1750 cm^3 of solution in order to convert it to a decimolar solution.

(2) 25.0 cm^3 *of a copper(II) sulphate solution were added to 20 cm^3 of potassium iodide solution, i.e., excess potassium iodide. The iodine liberated was titrated by 22.50 cm^3 of $0.1080 M$ sodium thiosulphate solution. Calculate the concentration of the solution in grams of the pentahydrate, $CuSO_4 \cdot 5H_2O$, per dm^3 of the solution.*

When a copper(II) salt is added to potassium iodide the copper(II) ions are reduced to form a precipitate of copper(I) iodide and the iodide ions are oxidized to iodine.

$$2Cu^{2+} + 4I^- \rightarrow 2CuI + I_2$$

1 mole of $Na_2S_2O_3$ reacts with 0.5 moles of I_2, which is formed from 1 mole of Cu^{2+} ions (or 1 mole of $CuSO_4 \cdot 5H_2O = 250$ g), i.e.

1000 cm^3 of 1 M $Na_2S_2O_3$ are equivalent to 250 g of $CuSO_4 \cdot 5H_2O$

22.50 cm^3 of $0.1080 M$ $Na_2S_2O_3$ are equivalent to

$$250 \times \frac{22.50}{1000} \times 0.1080 \text{ g}$$

This is the mass of the copper(II) salt in 25.0 cm^3, therefore

$$\text{the mass in 1 dm}^3 = 250 \times \frac{22.50}{1000} \times 0.1080 \times 40 = 24.3 \text{ g}$$

(3) *2.50 g of a sample of bleaching powder were ground with successive amounts of water, transferred to a measuring flask and made up to 250 cm^3 of mixture. 25.0 cm^3 of it, added to excess of potassium iodide solution and acidified, liberated iodine requiring 24.20 cm^3 of decimolar sodium thiosulphate solution. Calculate the percentage of 'available chlorine' in the bleaching powder.*

Available chlorine, i.e., the chlorine liberated by the action of dilute acid, produces iodine from potassium iodide solution by the reaction:

$$Cl_2 + 2I^- \rightarrow 2Cl^- + I_2$$

1 mole of $S_2O_3^{2-}$ reacts with 0.5 moles of I_2 which is formed from 0.5 moles of Cl_2 = 35.5 g

1000 cm³ of 1M $Na_2S_2O_3$ are equivalent to 35.5 g of Cl_2

24.20 cm³ of 0.1M $Na_2S_2O_3$ are equivalent to $35.5 \times \dfrac{24.20}{1000} \times 0.1$ g

This is the mass of chlorine formed from 25.0 cm³ of the mixture, therefore the mass from 250 cm³ is ten times greater and the percentage of available chlorine in the bleaching powder

$$= 35.5 \times \dfrac{24.20}{1000} \times 0.1 \times 10 \times \dfrac{100}{2.50} = 34.4$$

(4) *Iodate ions liberate iodine quantitatively from acidified potassium iodide solution. From the following data, calculate the number of moles of iodine (I_2) formed by reacting 1 mole of iodate ions with excess acidified potassium iodide solution. Suggest an equation for the reaction between iodate ions, iodide ions and hydrogen ions. 4.00 g of potassium iodate (KIO_3) were made up to 1 dm³ of aqueous solution. 25.0 cm³ of it were added to excess potassium iodide solution and then acidified. The liberated iodine required 28.0 cm³ of 0.1M sodium thiosulphate for reduction.*

Number of moles of KIO_3 (and hence IO_3^-) in 25.0 cm³ of solution

$$= 4.00 \times \dfrac{25.0}{1000} \times \dfrac{1}{214} = 0.0004673$$

Number of moles of $Na_2S_2O_3$ in 28.0 cm³ of 0.1M solution

$$= 0.1 \times \dfrac{28.0}{1000} = 0.0028$$

But 1 mole of $S_2O_3^{2-}$ combines with 0.5 moles of I_2, therefore the number of moles of iodine reacting with the thiosulphate = $0.0028 \times 0.5 = 0.0014$, i.e.,

0.0004673 moles of IO_3^- liberate 0.0014 of I_2

1 mole of IO_3^- liberates $\dfrac{0.0014}{0.0004673} = 2.996$ moles of I_2

That is, 1 mole of IO_3^- liberates 3 moles of I_2 and a likely equation for the reaction is

$$IO_3^- + 5I^- + 6H^+ \rightarrow 3I_2 + 3H_2O$$

Problems on Sodium Thiosulphate and Iodine

(Relative atomic masses will be found on page 132)

(22) A standardization of sodium thiosulphate solution yielded the following results. To an excess of acidified potassium iodide solution were added 25 cm^3 of 0.02M $KMnO_4$. The liberated iodine was found to need 24.2 cm^3 of the sodium thiosulphate solution. Calculate the molarity of the sodium thiosulphate solution.

(23) To about 2 g of potassium iodide dissolved in dilute sulphuric acid, 25 cm^3 of 0.02M $KMnO_4$ were added. 24.7 cm^3 of sodium thiosulphate solution were needed for titration of the liberated iodine. What is the concentration of the sodium thiosulphate solution in grams of hydrated salt per dm^3 and how much water must be added per dm^3 of the solution to make it 0.1M?

(24) 25 cm^3 of a solution of iodine in potassium iodide solution required 27.6 cm^3 of 0.0926M sodium thiosulphate solution to titrate the iodine. What is the mass of iodine per dm^3 of the solution?

(25) Calculate the molarity of a sodium thiosulphate solution from the following data of a standardization by copper: 1.500 g of pure copper foil was converted into 250 cm^3 of a solution of copper(II) nitrate free from nitric or nitrous acid and acidified with ethanoic acid. 25 cm^3 of this liberated sufficient iodine from potassium iodide solution to react with 24.6 cm^3 of the thiosulphate solution.

(26) In an experiment to find the percentage of copper in a sample of its basic carbonate, the following result was obtained: 3.100 g of the carbonate were converted into 250 cm^3 of copper(II) chloride solution free from mineral acid but acidified with ethanoic acid. 25 cm^3 of the solution liberated from potassium iodide solution an amount of iodine equivalent to 24.5 cm^3 of 0.1M sodium thiosulphate solution. Calculate the percentage of copper in the carbonate.

(27) 56 dm^3 of air containing a trace of chlorine were passed slowly through an excess of potassium iodide solution at 10°C and 100 000 N m^{-2} (750 mmHg). The solution was made up to 250 cm^3 with water and 25 cm^3 of it then required 20.6 cm^3 of 0.1M sodium thiosulphate solution to react with the iodine. What is the percentage by volume of chlorine in the air?

(28) 2.400 g of bleaching powder were made into a paste with water and diluted to 250 cm^3. After shaking 25 cm^3 were quickly drawn off and run into an excess of potassium iodide solution acidified with dilute ethanoic acid. The liberated iodine needed 22.0 cm^3 of 0.1M sodium thiosulphate solution. Calculate the percentage of available chlorine in the bleaching powder.

(29) An evaluation of manganese(IV) oxide in pyrolusite yielded the following figures: 1.400 g of the pyrolusite was warmed with concentrated hydrochloric acid and the chlorine liberated was passed through an excess of potassium iodide solution, after which the mixture was made up to 250 cm³. 25 cm³ of the solution required 24.4 cm³ of 0.098M sodium thiosulphate solution to titrate the iodine. Calculate the percentage of manganese(IV) oxide in the pyrolusite.

(30) 9.00 g of potassium iodide were dissolved in water and chlorine gas was passed through it. The resulting brown liquid was made up to 250 cm³ and 25 cm³ of it then required 21.1 cm³ of 0.1M sodium thiosulphate to react with the iodine. Calculate (a) what mass of chlorine was absorbed by the solution, (b) what mass of potassium iodide remained unchanged.

(31) Calculate the mass of potassium dichromate per dm³ of a solution from the following data: 25 cm³ of the solution were added to an excess of acidified potassium iodide solution. The liberated iodine needed 29.7 cm³ of 0.102M sodium thiosulphate solution.

(32) Calculate the mass of sodium sulphite as anhydrous salt per dm³ of solution from the following data: 25 cm³ of the solution were added slowly to 50 cm³ of 0.0512M iodine solution acidified with hydrochloric acid. The excess iodine required 17.7 cm³ of 0.0967M sodium thiosulphate.

(33) Calculate the concentration in g dm⁻³ of a solution of potassium iodate from the data: 25.0 cm³ of the potassium iodate solution were added to about 20 cm³ of a 10% solution of potassium iodide (i.e., excess iodide). On acidification, iodine was liberated and required 24.4 cm³ of 0.1M sodium thiosulphate solution to titrate it.

$$IO_3^- + 5I^- + 6H^+ \rightarrow 3I_2 + 3H_2O$$

(34) Bromate ions oxidize iodide ions according to the equation:

$$BrO_3^- + 6I^- + 6H^+ \rightarrow Br^- + 3H_2O + 3I_2$$

Bromate ions also oxidize hydrazine. Calculate the mole ratio of bromate ions (BrO_3^-) to hydrazine (N_2H_4) taking part in this reaction. 3.00 g of the salt $N_2H_5^+HSO_4^-$ were made up to 250 cm³ of aqueous solution. 25 cm³ of this solution were added to exactly 25 cm³ of 0.08M $KBrO_3$ solution (i.e., an excess) and then acidified. When the reaction was complete, potassium iodide solution was added and the liberated iodine required 27.7 cm³ of 0.1M sodium thiosulphate.

(35) When a solution of phenylamine (aniline) is added to a solution of bromine a white insoluble compound is formed. Use the following results to calculate the number of moles of bromine (Br_2) reacting with one mole of phenylamine ($C_6H_5NH_2$) and suggest a molecular formula for the white compound. 1.10 g of its salt phenylammonium chloride, $C_6H_5NH_3^+Cl^-$ were made up to 250 cm³ of aqueous solution. 25 cm³ of a 0.03M $KBrO_3$ solution were added to an excess of potassium bromide solution (thus producing a solution of

bromine according to the equation $BrO_3^- + 5Br^- + 6H^+ \rightarrow 3H_2O + 3Br_2$). 10 cm³ of the phenylamine solution were then run into the bromine solution. The excess unreacted bromine was then estimated by adding potassium iodide and titrating the liberated iodine with 0.1M $Na_2S_2O_3$ solution, 24.6 cm³ being required.

3

Silver Nitrate Titrations

Theory

In neutral solution, silver nitrate reacts with chlorides or bromides to precipitate silver chloride or silver bromide.

$$Ag^+(aq) + Cl^-(aq) \rightarrow AgCl(s), \quad Ag^+(aq) + Br^-(aq) \rightarrow AgBr(s)$$

That is, 1 mole of silver ions react completely with 1 mole of Cl^- ions or 1 mole of Br^- ions. Therefore, 1000 cm^3 of 1M $AgNO_3$ will react with 1 mole of Cl^- ions from, for example, 1 mole of NaCl or 0.5 moles of $CaCl_2$. The indicator usually used in silver nitrate titrations has been potassium chromate, the end point being the first permanent red–brown tinge in the mixture, caused by slight precipitation of silver chromate. Adsorption indicators, such as fluorescein, can also be used.

Examples

(1) 6.25 g *of common salt were made up to a* dm^3 *of solution in distilled water.* 25.0 cm^3 *of this solution required* 26.40 cm^3 *of decimolar silver nitrate solution. Calculate the percentage of sodium chloride in the sample of common salt.*

$$Ag^+ + Cl^- \rightarrow AgCl$$

From this, 1 mole of $AgNO_3$ will react with 1 mole of NaCl = 58.5 g, or 1000 cm^3 of 1M $AgNO_3$ will react with 58.5 g of NaCl.

To titrate the whole dm^3 of sodium chloride solution $(26.40 \times 40) \text{ cm}^3$ of decimolar silver nitrate solution would have been required.

$(26.40 \times 40) \text{ cm}^3$ of 0.1M $AgNO_3$ will react with

$$58.5 \times \frac{(26.40 \times 40)}{1000} \times 0.1 \text{ g NaCl}$$

This is the mass of sodium chloride in 1 dm³ of solution, therefore the percentage of sodium chloride in the sample

$$= 58.5 \times \frac{(26.40 \times 40)}{1000} \times 0.1 \times \frac{100}{6.25} = 98.9$$

(2) *A given solution contained potassium chloride and hydrochloric acid. 25.0 cm³ of it needed 24.80 cm³ of 0·0986M sodium hydroxide solution to neutralize the acid and the neutral solution then required 23.55 cm³ of 0.2M silver nitrate solution to precipitate all the chloride. Calculate the concentration in g dm^{-3} of potassium chloride and hydrochloric acid in the solution.*

The sodium hydroxide estimates the hydrochloric acid only.

$$HCl + NaOH \rightarrow NaCl + H_2O$$

From this, 1 mole of NaOH reacts with 1 mole of HCl = 36.5 g, or 1000 cm³ of 1M NaOH reacts with 36.5 g of HCl.

24.80 cm³ of 0.0986M NaOH reacts with $36.5 \times \dfrac{24.80}{1000} \times 0.0986$ g

This is the mass of HCl in 25 cm³, therefore the mass in 1 dm³

$$= 36.5 \times \frac{24.80}{1000} \times 0.0986 \times 40 = 3.57 \text{ g}$$

The silver nitrate estimates the total chloride present.

$$Ag^+ + Cl^- \rightarrow AgCl$$

1000 cm³ of 1M AgNO₃ reacts with 1 mole of Cl⁻ ions

23.55 cm³ of 0.2M AgNO₃ reacts with $1 \times \dfrac{23.55}{1000} \times 0.2 = 0.004710$,

i.e.,

the total number of moles of Cl⁻ in 25.0 cm³ = 0.004710

Number of moles of Cl⁻ from HCl in 25.0 cm³ = $0.0986 \times \dfrac{24.80}{1000}$

$$= 0.002445$$

Number of moles of Cl⁻ from KCl in 25.0 cm³ = 0.004710
$$- 0.002445$$
$$= 0.002265$$

1 mole of KCl is 74.5 g, therefore the mass of KCl in 1 dm³

$$= 0.002265 \times 40 \times 74.5 = 0.75 \text{ g}$$

(3) *6.150 g of barium chloride crystals, $BaCl_2 \cdot xH_2O$, were made up to 500 cm³ of solution in distilled water. 25.0 cm³ of this solution were treated* with a moderate excess of sodium sulphate solution, after which 25.22 cm³ of decimolar silver nitrate solution were needed to titrate the chloride in solution. Calculate x.

The sodium sulphate merely serves to precipitate the Ba^{2+} ions as $BaSO_4$ which prevents them reacting with the chromate ions of the indicator.

1000 cm³ of 1M $AgNO_3$ reacts with 1 mole of Cl^- (i.e., 0.5 moles of $BaCl_2.xH_2O$)

From the titration,

25.22 cm³ of 0.1M $AgNO_3$ reacts with $\dfrac{6.15}{20}$ g of $BaCl_2.xH_2O$

1000 cm³ of 1M $AgNO_3$ reacts with $\dfrac{6.15}{20} \times \dfrac{1000}{25.22} \times 10 = 121.9$ g

This is 0.5 moles of $BaCl_2.xH_2O$, therefore

$$208 + 18x = 2 \times 121.9$$
$$x = 2$$

The crystalline salt has the formula $BaCl_2.2H_2O$.

(4) 1.16 g of the chloride of aluminium were made up to 250 cm³ of aqueous solution. 25 cm³ of this solution, using dichlorofluorescein as indicator, required 26.0 cm³ of 0.1M $AgNO_3$. Calculate the number of moles of chloride combined with 1 mole of aluminium and hence the simplest formula of the chloride.

$$Ag^+ + Cl^- \rightarrow AgCl$$

1000 cm³ of 1M $AgNO_3$ reacts with 1 mole of Cl^- ions

26.0 cm³ of 0.1M $AgNO_3$ reacts with $1 \times \dfrac{26.0}{1000} \times 0.1$
= 0.0026 moles

The mass of chloride in 25 cm³ $= 0.0026 \times 35.5 = 0.0923$ g
The mass of aluminium chloride in 25 cm³ $= 0.116$ g
and therefore, the mass of aluminium in 25 cm³ of solution

$$= 0.0116 - 0.0923 = 0.0237 \text{ g}$$

i.e., in 25 cm³ of solution 0.0237 g of Al are combined with 0.0026 moles of Cl

1 mole of Al (27 g) is combined with $0.0026 \times \dfrac{27}{0.0237}$
= 2.963 moles of Cl

Therefore the simplest formula of chloride is $AlCl_3$.

Problems on Silver Nitrate

(Relative atomic masses will be found on page 132)

(1) Calculate the percentage by mass of chlorine in a sample of ammonium chloride from the following results: 5.00 g of ammonium chloride were made up to 1 dm³ of aqueous solution. 25 cm³ of this required 23.0 cm³ of 0.1M $AgNO_3$ for complete precipitation.

(2) 1.500 g of an impure sample of common salt was made up to 250 cm³ of aqueous solution. 25 cm³ of this solution needed 24.6 cm³ of decimolar silver nitrate solution for complete precipitation. Calculate the percentage by mass of sodium chloride in the sample.

(3) 25 cm³ of hydrochloric acid were neutralized by chloride free chalk. The solution then required, for precipitation, 32.0 cm³ of a solution of silver nitrate of which 23.0 cm³ had been required to react with 25 cm³ of a solution containing 5.00 g of potassium chloride per dm³. Calculate the concentration of the acid in g dm^{-3}.

(4) Calculate the percentage by mass of sodium chloride in a mixture of this compound with potassium chloride from the following data: 1.000 g of the mixture was made up to 100 cm³ of aqueous solution and 10 cm³ of this solution required 16.5 cm³ of 0.1M $AgNO_3$ for precipitation.

(5) 25 cm³ of a solution containing potassium chloride and hydrochloric acid required 18.2 cm³ of 0.1M KOH for neutralization. To another 25 cm³ of the solution was added excess of chloride free chalk after which the solution required 20.2 cm³ of 0.2M $AgNO_3$ for precipitation. Calculate the masses of hydrochloric acid and potassium chloride per dm³ of solution.

(6) 9.500 g of a mixture of potassium chloride and potassium bromide were made up to 1 dm³ of aqueous solution. 25 cm³ of this solution required 24.8 cm³ of 0.103M $AgNO_3$ solution for precipitation. Calculate the percentage by mass of each salt in the mixture.

(7) Calculate x in the formula $BaCl_2 \cdot xH_2O$ from the following: 3.05 g of barium chloride crystals were made up to 250 cm³ with distilled water. To 25 cm³ of this solution was added excess potassium sulphate, after which 23.5 cm³ of 0.106M $AgNO_3$ were required to precipitate the chloride.

(8) 1.000 g of a mixture of potassium chloride and anhydrous potassium carbonate was made up to 100 cm³ of aqueous solution. 25 cm³ of this solution required 12.5 cm³ of 0.1M HCl for neutralization with methyl orange indicator. How many cm³ of 0.05M $AgNO_3$ would be required to precipitate the chloride from the neutral solution?

(9) 1.000 g of a mixture of anhydrous sodium and potassium carbonates was made up to 250 cm³ of aqueous solution. 25 cm³ of this

solution were neutralized, with methyl orange as indicator, by 20.0 cm^3 of hydrochloric acid of unknown concentration. The neutral solution then required 16.24 cm^3 of 0.1M AgNO$_3$ for precipitation. Calculate (*a*) the percentage by mass of potassium carbonate in the mixture, (*b*) the concentration of the hydrochloric acid in g dm^{-3}.

(10) To 100 cm^3 of well water was added potassium chromate as indicator, and then a silver nitrate solution containing 4.79 g of silver nitrate per dm^3. 4.60 cm^3 of the silver nitrate solution produced a permanent red–brown tinge. What is the number of parts by mass of chlorine in 1 000 000 parts by mass of the well water?

(11) 5.000 g of potassium chlorate were heated until evolution of oxygen had ceased and the residue was made up to 500 cm^3 of aqueous solution. 25 cm^3 of this solution required 20.4 cm^3 of 0.1M AgNO$_3$ for precipitation. Calculate the percentage of chlorine in the potassium chlorate (KClO$_3$).

(12) 0.500 g of a mixture of dilute sulphuric and hydrochloric acids was made up to 100 cm^3 of solution with distilled water. 20 cm^3 of this solution required 17.49 cm^3 of 0.1M NaOH for neutralization and this neutral solution needed 10.96 cm^3 of 0.1M AgNO$_3$ to precipitate the chloride. Calculate the ratio, water : hydrogen chloride : sulphuric acid, by mass, in the original mixture.

(13) 1.00 g of a chloride of phosphorus was dissolved in water and made up to 250 cm^3 of solution. 25 cm^3 of this solution, after neutralization, required 23.9 cm^3 of 0.1M AgNO$_3$ for complete precipitation of the chloride. Calculate the number of moles of chloride combined with one mole of phosphorus and hence deduce the simplest formula of the chloride.

(14) 1.52 g of a chloride of type MCl$_2$ were made up to 250 cm^3 of aqueous solution. 25 cm^3 of this solution required 27.3 cm^3 of 0.1M AgNO$_3$ for complete precipitation of the chloride. Calculate the mass of 1 mole of MCl$_2$ and hence the relative atomic mass of M.

4

Gravimetric Analysis

Examples

Methods of gravimetric analysis are quite varied. The following are typical calculations relating to some of the simpler operations.

(1) 0.915 g *of slightly impure barium chloride crystals, $BaCl_2 \cdot xH_2O$, were dissolved in hot water and slight excess of silver nitrate solution was added. The precipitated silver chloride, after filtration, washing and drying at about* $120°C$, *was found to have a mass of* 1.077 g. *Calculate the probable value of* x.

This estimation depends on the reaction:
$$BaCl_2 + 2AgNO_3 \rightarrow 2AgCl + Ba(NO_3)_2$$
From this 1 mole of $BaCl_2 \cdot xH_2O$ will produce 2 moles of AgCl, and
$$\frac{BaCl_2 \cdot xH_2O}{2AgCl} = \frac{0.915}{1.077}$$
Inserting the appropriate masses,
$$\frac{208 + 18x}{2 \times 143.5} = \frac{0.915}{1.077}$$
Simplifying this, $208 + 18x = 243.8$ and
$$x = \frac{35.8}{18} = 1.99$$
i.e., the probable value of x is 2.

(2) 0.936 g *of a sample of hydrated copper(II) sulphate crystals was dissolved in hot distilled water. Careful addition of sodium hydroxide solution precipitated copper(II) oxide, which, when filtered, washed and dried to constant mass, was found to have a mass of* 0.298 g. *Calculate the percentage of copper in the crystals and compare the result with the calculated figure for* $CuSO_4 \cdot 5H_2O$.

The reaction used is:
$$CuSO_4 + 2NaOH \rightarrow Na_2SO_4 + H_2O + CuO$$
The experimental percentage of copper in the crystals
$$= 0.298 \times \frac{64}{80} \times \frac{100}{0.936} = 25.5$$
The calculated percentage
$$\frac{Cu}{CuSO_4 \cdot 5H_2O} \times 100 = \frac{64}{250} \times 100 = 25.6$$

(3) 1.046 g *of the hydrated chloride of a certain metal were dissolved in water and silver chloride was precipitated by the addition of a slight excess of silver nitrate solution. After purification the silver chloride had a mass of 1.231 g. Calculate the percentage of chlorine in the hydrate. If it is of the form, $MCl_2 \cdot 2H_2O$, calculate the relative atomic mass of the metal, M.*

The mass of chlorine in the silver chloride precipitate is $1.231 \times \dfrac{35.5}{143.5}$ g.

Consequently, the percentage of chlorine in the hydrate
$$= 1.231 \times \frac{35.5}{143.5} \times \frac{100}{1.046} = 29.1$$
From the formula given, the percentage of chlorine is
$$\frac{2Cl}{MCl_2 \cdot 2H_2O} \times 100 \quad \text{which is equal to} \quad \frac{71}{A + 107} \times 100 = 29.1$$
Solving this for A gives $A = 137.0$, which is the required relative atomic mass.

Problems on Gravimetric Analysis
(Relative atomic masses will be found on page 132)

(1) 1.362 g of barium chloride crystals were dissolved in distilled water and heated. Ammonium chloride solution and excess boiling dilute sulphuric acid were added. The precipitate was filtered, washed and dried. From the following results calculate the percentage of barium in the barium chloride crystals.

Mass of Gooch crucible . . . . . 10.732 g
Mass of Gooch crucible and barium sulphate . 12.033 g

(2) 1.523 g of a mixture of sodium chloride and potassium nitrate was dissolved in distilled water and dilute nitric acid and silver nitrate solution were added until precipitation was complete. The residue was filtered off, washed and dried, and was found to have a mass of 2.943 g. Find the mass of potassium nitrate in the original mixture.

(3) 0.887 g of a mixture of NaCl and KCl yielded 1.913 g of silver chloride. Calculate the percentage by mass of each salt in the mixture.

(4) 2.456 g of a mixture of hydrated aluminium potassium sulphate (potash alum) and anhydrous potassium sulphate were dissolved in a small quantity of water and heated. Ammonium chloride and ammonia were added to complete the precipitation of aluminium hydroxide. The precipitate was filtered off, washed and heated with a blowpipe and 0.170 g of aluminium oxide was obtained. What was the percentage by mass of anhydrous potassium sulphate in the mixture?

(5) 0.915 g of barium chloride crystals was dissolved in water and the barium was precipitated by the addition of ammonium chromate. The purified and dried barium chromate had a mass of 0.9847 g. Calculate the percentage by mass of water in the sample of crystals.

(6) Calculate the relative atomic mass of a divalent metal X from the following data. 2.370 g of the anhydrous metal sulphate were dissolved in water, a little dilute sulphuric acid added and the mixture boiled. Sodium hydroxide solution was added until a few drops produced no further effect. The residue, which was the normal oxide, was filtered off, washed and dried. It was found to have a mass of 1.185 g.

(7) Find the number of moles of water of crystallization in one mole of magnesium sulphate crystals from the following experiment: 1.227 g of the crystals were dissolved in water, and ammonium chloride and ammonia solution added and the solution was boiled. Excess of disodium hydrogenphosphate(V) solution was added, and the precipitated ammonium magnesium phosphate(V) heated for some time until the conversion to magnesium heptaoxodiphosphate(V), $Mg_2P_2O_7$, was complete. This residue was found to have a mass of 0.5538 g.

(8) Calculate the percentage of iron in a sample of iron ore from the following experimental results. 3.720 g of haematite (Fe_2O_3) were dissolved in warm concentrated hydrochloric acid and carefully diluted to 250 cm³ of solution. 25 cm³ of this iron(III) chloride solution were treated as follows. The iron was precipitated as iron(III) hydroxide by the addition of excess ammonium hydroxide and the precipitate was filtered, washed and ignited to iron(III) oxide.

Mass of crucible	15.2861 g
Mass of crucible after 1st ignition	15.6209 g
Mass of crucible after 2nd ignition	15.6205 g
Mass of crucible after 3rd ignition	15.6205 g

(9) Calculate x in the formula $CuSO_4 \cdot xH_2O$ from the results of the following experiment. 0.6238 g of the copper(II) sulphate crystals was dissolved in water and black copper(II) oxide was precipitated by treatment with boiling sodium hydroxide solution. After complete precipitation of the oxide, the latter was purified and dried. It was then found to have a mass of 0.1996 g.

5

Colligative Properties Not Involving Dissociation or Association

The vapour pressure, boiling point, freezing point and osmotic pressure of a dilute solution of a non-volatile solute depend on the number of solute particles present in a fixed amount of solvent and hence are called colligative properties.

LOWERING OF VAPOUR PRESSURE

Theory

Dissolution of a non-volatile solute in a solvent lowers the vapour pressure of the solvent. Relative molecular masses of solutes can be determined by direct observation of vapour pressure changes. The experimental law governing the vapour pressures of dilute solutions of non-volatile, non-electrolytes was formulated by Raoult and it is:

$$\frac{p_0 - p_1}{p_0} = \frac{n}{n + N}$$

where p_0 is the vapour pressure of the pure solvent ⎫ at constant
p_1 is the vapour pressure of a dilute solution ⎬ temperature
n is the number of moles of solute
N is the number of moles of solvent.

Example

The vapour pressure of ethoxyethane at a certain temperature is 664 300 $N\,m^{-2}$. When 3.6 g of organic acid were dissolved in 67 g of ethoxyethane at that temperature, the vapour pressure fell to 629 600 $N\,m^{-2}$. Calculate the relative molecular mass of the acid. (Ethoxyethane is $C_4H_{10}O$.)

Applying the above formula,

$$\frac{664\,300 - 629\,600}{664\,300} = \frac{\dfrac{3.6}{M}}{\dfrac{3.6}{M} + \dfrac{67}{74}}$$

where M is the required relative molecular mass. Solving this equation, $M = 72.1$.

An approximate form of the equation can be used if the solution is very dilute, because, in that case, the number of moles, n, of the solute is negligible in comparison with the number of moles, N, of solvent. The approximate form is:

$$\frac{p_0 - p_1}{p_0} = \frac{n}{N}$$

Applied to the above case, this approximation gives:

$$\frac{664\,300 - 629\,600}{664\,300} = \frac{\dfrac{3.6}{M}}{\dfrac{67}{74}}$$

This yields the result, $M = 76$

The approximation should **not** be used unless the solution is **very** dilute.

BOILING POINT ELEVATION AND FREEZING POINT DEPRESSION

Theory

The lowering of the vapour pressure of a solvent by a non-volatile solute is reflected in an elevation of the boiling point and a depression of the freezing point of the solvent.

The laws governing these phenomena are:

(1) At constant pressure and for dilute solutions of a given solute, the elevation of boiling point, or the depression of freezing point, of the solution is directly proportional to its mass concentration.

(2) Dilute solutions of all covalent solutes, with the same molar concentrations, have the same boiling points (or freezing points), at the same pressure, in the same solvent.

For reference, the data on boiling point and freezing point are expressed in the form of a boiling point constant or freezing point constant, K, which is characteristic of the solvent in question. This constant is the elevation of the boiling point, or the depression of the freezing point (in °C), that would be produced by dissolving one mole of any covalent solute in 1000 g of the solvent, assuming that the laws of dilute solutions still apply. In the past, the reference amount of 100 g of solvent has also been used, in which case the constant is ten times greater than for 1000 g.

The constant can be found: (i) by a formula derived from thermodynamical considerations, (ii) from experimental observation.

(i) The formula is:

$$K = \frac{RT^2}{L}$$

where R is the molar gas constant, T the boiling point, or freezing point, of the solvent at atmospheric pressure (760 mmHg or 101.3 kN m^{-2}) on the Kelvin scale, and L is the latent heat of vaporization, or fusion, of the solvent in J kg^{-1} at the boiling point or freezing point.

R is equal to 8.31 J K^{-1} mol^{-1}, and the latent heat of fusion of water is 334 J g^{-1}. Therefore, the freezing point constant for water is:

$$K = \frac{8.31 \times (273)^2}{1000 \times 334} = 1.86°C$$

(ii) The constant can also be found from experimental observations, using a solute of known relative molecular mass. It is given by the formula:

$$K = t \times \frac{M}{m} \times \frac{w}{1000}$$

where t °C is the boiling point elevation, or freezing point depression, produced in w grams of the solvent by m g of solute of relative molecular mass M. The formula is the direct expression of the laws stated earlier.

An experiment shows that 0.90 g of urea (relative molecular mass = 60) depresses the freezing point of 30.0 g of water by 0.93°C. From this:

$$K = 0.93 \times \frac{60}{0.90} \times \frac{30.0}{1000} = 1.86°C$$

Alternatively, working from first principles, 0.90 g of urea dissolved in 30.0 g of water causes a depression of 0.93°C, so 60 g of urea dissolved in 30.0 g of water cause a depression of

$$0.93 \times \frac{60}{0.90} °C$$

and 60 g of urea dissolved in 1000 g of water cause a depression of

$$0.93 \times \frac{60}{0.90} \times \frac{30.0}{1000} = 1.86°C$$

1.86°C is the depression constant as it is the depression caused by one mole of urea dissolved in 1000 g of water.

If K is known for a given solvent, the molecular mass of a given solute can be expressed by the rearrangement:

$$M = \frac{m \times K \times 1000}{t \times w}$$

Thus a relative molecular mass can be calculated after suitable experimental observations have been made. Alternatively (as illustrated in the following examples), the relative molecular mass can be calculated by working from first principles.

Examples

(1) *1.50 g of certain compound, when dissolved in 30 g of water, produced a solution freezing at $-1.04°C$, at 101 300 $N\,m^{-2}$ pressure. Calculate the relative molecular mass of the compound. ($K = 1.86°C$ per 1000 g of water.)*

Using the formula above,

$$\text{Relative molecular mass} = \frac{1.50 \times 1.86 \times 1000}{1.04 \times 30} = 89.4$$

If working from first principles is preferred, the calculation can be stated as:

1.04°C is the depression in 30 g of water by dissolving 1.50 g of solute, so 1.86°C is the depression in 1000 g of water by dissolving

$$1.50 \times \frac{1.86}{1.04} \times \frac{1000}{30} \text{ g of solute}$$

the first fraction represents the fact that the mass of solute must increase to produce the greater f.p. depression; the second fraction represents the increase in mass of solute necessary to compensate for the increased mass of solvent. The complete fraction is the same as that obtained from the formula above and is, by definition of the constant, the molar mass of the solute.

(2) *What mass of cane sugar (relative molecular mass 342) must be added to 105 g of water to produce a solution boiling at 100.060°C, at 101.3 kN m^{-2} pressure? (K = 0.52°C per 1000 g of water.)*

Let the required mass of cane sugar be m g. Using the formula for relative molecular mass above,

$$342 = \frac{m \times 0.52 \times 1000}{0.060 \times 105}$$

From this, $m = \dfrac{342 \times 0.060 \times 105}{0.52 \times 1000} = 4.14$

Alternatively, working from first principles,

0.52°C is the elevation in 1000 g of water by 342 g of cane sugar,

so

0.060°C is the elevation in 105 g of water by

$$342 \times \frac{0.060}{0.52} \times \frac{105}{1000} \text{ g of cane sugar}$$

For the significance of these fractions, see comments on Example 1 above. The final result is the same as by the formula.

(3) *1.20 g of iodine raised the boiling point of a certain solvent by 0.400°C. If the volume of solvent used was 32.0 cm^3, calculate the molecular state of the iodine in it. (For the solvent, K = 2.1°C per 1000 g and density = 0.75 g cm^{-3}.*

Using the formula, M being the relative molecular mass of iodine,

$$M = \frac{1.20 \times 2.1 \times 1000}{0.400 \times (32.0 \times 0.75)} = 262$$

Thus the molecular state is diatomic, I_2.

If working from first principles is preferred,

0.0400°C is the elevation in (32.0 × 0.75) g of solvent by 1.20 g of iodine,

so

2.10°C is the elevation in 1000 g of solvent by

$$1.20 \times \frac{2.10}{0.400} \times \frac{1000}{(32.0 \times 0.75)} \text{ g of iodine}$$

For the significance of these fractions, see comments on Example 1. The total fraction is the same as by the use of the formula above.

OSMOTIC PRESSURE

Theory

The laws of osmotic pressure can be summarized in the following way:

(1) For dilute solutions of a given solute, at constant temperature, the osmotic pressure of the solution is directly proportional to its mass concentration (which means that it is inversely proportional to the volume of solution containing a fixed mass of solute).
(2) The osmotic pressure of given solution is directly proportional to its temperature on the Kelvin scale.
(3) Dilute solutions of different covalent solutes, with the same molar concentration, have the same osmotic pressure at the same temperature.

These laws are similar to those expressing the behaviour of gases with respect to temperature and pressure changes and give rise to a similar mathematical representation, as:

For a given mass of gas	*For a given mass of covalent solute in dilute solution*
$\dfrac{PV}{T}$ = a constant	$\dfrac{\Pi \times V}{T}$ = a constant
	(where Π = osmotic pressure)

If referred to 1 mole of gas or solute respectively, these constants are quantitatively identical and are known as the molar gas constant, R. That is, with reference to 1 mole of gas or 1 mole of covalent solute in dilute solution,

$$PV = RT \quad \text{or} \quad \Pi V = RT$$

This means that, just as 1 mole of any gas in $22.4\,dm^3$ of volume at $0°C$ exerts a pressure of 1 atmosphere, so 1 mole of any covalent solute in $22.4\,dm^3$ of solution at $0°C$ exerts an osmotic pressure of one atmosphere ($760\,mmHg$ or $101\,300\,N\,m^{-2}$). R is $0.082\,atm\,dm^3\,K^{-1}\,mol^{-1}$ or $8.31\,J\,K^{-1}\,mol^{-1}$.

In general, calculations relating to osmotic pressure are merely applications of the above facts in appropriate forms. The formula given above (right) is not always convenient for osmotic pressure calculations for, when solutions are in question, concentration, c, is a more convenient factor to deal with than the volume containing one mole of solute. Consequently, the osmotic formula is often used in the form:

$$\frac{\Pi}{c \times T} = \text{a constant}$$

c appears in the denominator because, for a given mass of solute, c is inversely proportional to V.

The SI unit for pressure is $N\,m^{-2}$ and this should be used in accurate work but the units mmHg and atmospheres will be found in many problems. Examples and problems using each system of units are given below.

Examples

(1) *Calculate the osmotic pressure in $N\,m^{-2}$ of a solution containing 2.20 g of urea (relative molecular mass 60) in $100\,cm^3$ at $15°C$.*

Using the formula quoted above,

$$\frac{\Pi_1}{c_1 \times T_1} = \frac{\Pi_2}{c_2 \times T_2}$$

Inserting the given experimental data on the left,

$$\frac{\Pi_1}{\frac{2.20}{100} \times 288} = \frac{101\,300}{\frac{60}{22\,400} \times 273}$$

where Π_1 is the required osmotic pressure of the urea solution. The right-hand side of the equation states the fact that one mole (60 g) of urea exerts 1 atmosphere ($101\,300\,N\,m^{-2}$) of osmotic pressure if in $22.4\,dm^3$ of solution at $0°C$.

From this equation,

$$\Pi_1 = \frac{2.20}{100} \times 288 \times \frac{22\,400}{60 \times 273} \times 101\,300$$
$$= 877\,000\,N\,m^{-2}$$

Alternatively, the calculation can be conducted from first principles by expressing each change involved as a fraction which states the requirements of the appropriate law. For example, the above calculation would take the form:

60 g of urea in $22\,400\,cm^3$ of solution at $0°C$ exerts $101\,300\,N\,m^{-2}$

2.20 g of urea in $100\,cm^3$ of solution at $15°C$ give an osmotic pressure of

$$101\,300 \times \frac{2.20}{60} \times \frac{22\,400}{100} \times \frac{288}{273} \text{N m}^{-2}$$

The first fraction expresses the fact that the Π decreases as the mass of urea falls from 60 g to 2.20 g; the second fraction expresses the fact that the Π increases as the volume of the solution decreases from 22 400 cm³ to 100 cm³ and the third fraction states the fact that the Π increases as temperature rises. The resulting expression is the same as the one obtained from the use of the formula above.

A third method of solving this problem is to use the equation,

$$\Pi V = RT$$

$R = 0.082\,\text{atm dm}^3\,\text{K}^{-1}\,\text{mol}^{-1}$ and V is the volume of solution, in dm³, containing one mole of solute, i.e., $V = \frac{0.100}{2.20} \times 60\,\text{dm}^3$. Substituting into the equation,

$$\Pi \times \frac{0.100}{2.20} \times 60 = 0.082 \times 288$$

From this,

$$\Pi = \frac{0.082 \times 288 \times 2.20}{0.100 \times 60} \text{ atmospheres}$$

$$\Pi = \frac{0.082 \times 288 \times 2.20}{0.100 \times 60} \times 101\,300\,\text{N m}^{-2}$$

$$\Pi = 877\,000\,\text{N m}^{-2}$$

(2) *A solution containing 1.50 g of a covalent solute in 1500 cm³ of solution at 10°C has an osmotic pressure of 320 mmHg. Calculate the relative molecular mass of the solute.*

Using the formula and inserting the experimental data on the left-hand side

$$\frac{\Pi_1}{c_1 \times T_1} = \frac{\Pi_2}{c_2 \times T_2}$$

$$\frac{320}{\frac{1.50}{1500} \times 283} = \frac{760}{\frac{M}{22\,400} \times 273}$$

where M is the required relative molecular mass. From this

$$M = \frac{1.50}{1500} \times 283 \times 760 \times \frac{22\,400}{320 \times 273} = 55$$

For comments on this method, see Example 1.

Alternatively, working from first principles,

320 mmHg is the Π given in 1500 cm^3 of solution at 10°C by 1.5 g of solute,

so

760 mmHg is the osmotic pressure given in 22 400 cm^3 of solution at 0°C by

$$1.5 \times \frac{760}{320} \times \frac{22\,400}{1500} \times \frac{283}{273} \text{ g of solute}$$

The first fraction expresses the fact that the mass of solute must increase to provide the higher osmotic pressure; the second fraction expresses the fact that the mass of solute increases as the volume of solution increases, to keep the same osmotic pressure; the third fraction expresses the fact that the mass of solute must be increased to maintain the same osmotic pressure as the temperature falls. The complete fraction is the same as that obtained from the use of the formula above.

(3) *Calculate the concentration in g dm^{-3} of a glucose solution which has an osmotic pressure of 2.32 atmospheres at 12°C. (Glucose is $C_6H_{12}O_6$.)*

Let the concentration of the solution be x g dm^{-3}. Using the formula and inserting the experimental figures on the left-hand side,

$$\frac{\Pi_1}{c_1 \times T_1} = \frac{\Pi_2}{c_2 \times T_2}$$

$$\frac{2.32}{\frac{x}{1} \times 285} = \frac{1}{\frac{180}{22.4} \times 273}$$

From this

$$x = 2.32 \times \frac{180}{22.4} \times 273 \times \frac{1}{285} = 17.9 \text{ g}$$

i.e., the concentration of the glucose solution is 17.9 g dm^{-3}. For comments on this method see Example 1.

Alternatively, working from first principles,

1 atm of osmotic pressure is given at 0°C in 22.4 dm^3 of solution by 180 g of glucose,

so

2.32 atm of osmotic pressure is given at 12°C in 1 dm^3 of solution by

$$180 \times \frac{2.32}{1} \times \frac{273}{285} \times \frac{1}{22.4} \text{ g of glucose}$$

This is the same fraction as the one obtained by the formula above; for comments on this method see Examples 1 and 2.

Problems on Colligative Properties Not Involving Dissociation or Association

(*Relative atomic masses will be found on page* 132)

VAPOUR PRESSURE

(1) The vapour pressure of propanone at 30°C is 37 330 N m^{-2}. When 6.0 g of a non-volatile organic compound are dissolved in 120 g of propanone the vapour pressure is reduced to 35 570 N m^{-2}. Calculate the relative molecular mass of the compound.

(2) The vapour pressure of benzene at 20°C is 10 000 N m^{-2}. What would be the effect on this vapour pressure of dissolving 4.1 g of naphthalene ($C_{10}H_8$) in 60 g of benzene, at the same temperature?

(3) The vapour pressure of a solution of sucrose, $C_{12}H_{22}O_{11}$, in water has a vapour pressure of 2311 N m^{-2} at 20°C. Pure water at this temperature has a vapour pressure of 2333 N m^{-2}. What is the concentration of this sucrose solution in g dm^{-3}.

BOILING POINT AND FREEZING POINT

(4) 0.60 g of urea (CON_2H_4) depresses the freezing point of 30.00 g of a certain solvent by 0.62°C. Calculate the depression of freezing point constant for the solvent.

(5) What mass of cane sugar ($C_{12}H_{22}O_{11}$) must be added to 35 g of water to raise the boiling point of the solution at 760 mmHg pressure to 100.015°C? ($K = 0.52$°C per 1000 g of water.)

(6) What is the freezing point at 101 300 N m^{-2} pressure of a solution containing 2.00 g of urea in 75 g of water? (Urea CON_2H_4.) ($K = 1.86$°C per 1000 g of water.)

(7) The boiling point of benzene under certain pressure conditions is 80.000°C. Calculate the boiling point of a solution containing 5 g of 2,4,6-trinitrophenol, $HOC_6H_2(NO_2)_3$ (picric acid) in 100 g of benzene, under these pressure conditions. ($K = 2.6$°C per 1000 g of benzene.)

(8) Calculate the relative molecular mass of a compound 2% aqueous solution of which boils at 99.877°C when the boiling of pure water is 99.700°C. ($K = 0.52$°C per 1000 g of water.)

(9) Calculate the relative molecular mass of a compound of which, when dissolved in 654 g of ethanoic acid, freezing at 16.62°C. (The freezing point of pure 16.75°C; $K = 3.9$°C per 1000 g of ethanoic acid

(10) 0.40 g of camphor when dissolved in 33.5 g of trichloromethane (chloroform) produces a solution boiling at 0.30°C above the boiling point of the pure solvent. Calculate the relative molecular mass of the camphor. ($K = 3.9$°C per 1000 g of trichloromethane.)

(11) 0.72 g of iodine raises the boiling point of 20.00 g of ethoxyethane by 0.30°C. Calculate the relative molecular mass of iodine in ethoxyethane. ($K = 2.1$°C per 1000 g of ethoxyethane.)

(12) 2.796 g of ethanamide raised the boiling point from 100.40°C to 100.75°C when dissolved in 70.4 g of water. Calculate the relative molecular mass of ethanamide. ($K = 0.52$°C per 1000 g of water.)

(13) What is the relative molecular mass of a compound if a 2% solution of it in benzene freezes at 4.291°C? ($K = 5.0$°C per 1000 g of benzene. Freezing point of pure benzene = 5.400°C.)

(14) 1.00 g of compound, when dissolved in 20.4 g of water, produced a solution freezing at -1.05°C at 101 300 N m^{-2} (760 mmHg) pressure. Calculate the relative molecular mass of the compound. ($K = 1.86$°C per 1000 g of water.)

(15) Calculate the relative molecular mass of benzoic acid on the basis of the following experimental results.

Mass of ethanol 16.150 g.

Boiling point/°C	Mass of benzoic acid dissolved/g
78.100	—
78.230	0.2400
78.360	0.4727
78.505	0.7470

$K = 1.07$°C per 1000 g of ethanol

(16) 2.00 g of phosphorus raise the boiling point of 37.4 g of carbon disulphide by 1.003°C. Show that the molecular formula of phosphorus in carbon disulphide is P_4. ($K = 2.35$°C per 1000 g of carbon disulphide.)

(17) A solution of 0.36 g of sulphur in 24.0 g of carbon disulphide boils at 0.14°C above the boiling point of the pure solvent at the same pressure. What is the molecular formula of sulphur in this solvent? ($K = 2.35$°C per 1000 g of carbon disulphide.)

(18) A motor car radiator has a capacity of 4 dm^3. What is the least mass of glycerol, $C_3H_8O_3$, which will convert the water it contains when full into a non-freezing mixture, if it is to encounter temperatures as low as -6°C. ($K = 1.86$°C per 1000 g of water. Assume the laws which hold for dilute solutions are valid for concentrated ones and that the solution has a density of 1 kg dm^{-3}.)

OSMOTIC PRESSURE

(19) A certain solution of urea has an osmotic pressure of 100 000 Nm^{-2} at 15°C. What will be its osmotic pressure at 0°C?

(20) An aqueous solution of a certain compound, containing 7.00 g of the compound in 100 cm³ of solution, had an osmotic pressure of 9.30 atmospheres at 18°C. What mass of this compound in 250 cm³ of solution would give an osmotic pressure of 10.00 atmospheres at 10°C?

(21) An aqueous solution containing 1.00 g of cane sugar in 200 cm³ of solution gave an osmotic pressure of 249 mmHg at 0°C. Calculate the relative molecular mass of the cane sugar.

(22) A solution of a certain sugar of relative molecular mass 342 has an osmotic pressure at 12°C of 127 600 Nm^{-2}. What is the concentration of the solution in grams of the sugar per dm³?

(23) 3.50 g of glucose were dissolved to make 1 dm³ of aqueous solution and the solution was found to have an osmotic pressure of 0.462 atm at 17°C. Calculate the relative molecular mass of the glucose. What would be the osmotic pressure of this solution at 35°C?

(24) Calculate the osmotic pressure in Nm^{-2} of a solution of urea (CON_2H_4) at 42°C, which contains 20 g dm⁻³.

(25) An aqueous solution of 1.00 g of a very weak acid in 1 dm³ of solution has an osmotic pressure of 288 mmHg at 12°C. Calculate the relative molecular mass of the acid. At what temperature would you expect this solution to freeze? (K = 1.86°C per 1000 g of water.)

(26) Calculate the relative molecular mass of glycerol from the following data: A 2% solution of glycerol has an osmotic pressure of 5.30 atm at 25°C.

(27) A solution of urea, CON_2H_4, containing 1.754 g dm⁻³, is isotonic at the same temperature with a solution of 10.00 g of certain sugar in 1 dm³ of aqueous solution. Calculate the relative molecular mass of the sugar.

(28) What mass of phenol, C_6H_5OH, must be dissolved in 25 cm³ of water to give an aqueous solution having an osmotic pressure of one atmosphere at 21°C?

(29) What mass of urea of relative molecular mass 60 is present per dm³ of an aqueous solution which has an osmotic pressure of 430 400 Nm^{-2} at 14°C?

(30) What will be the ratio of the masses of glucose and glycerol which must be dissolved to give 1 dm³ of aqueous solution of the same osmotic pressure at 27°C? (Glycerol $C_3H_8O_3$; glucose $C_6H_{12}O_6$.)

(31) Two solutions, A and B, of glucose have the same osmotic pressure of $324\,100\,N\,m^{-2}$ at 0°C and 20°C respectively. Calculate how many more grams of glucose, $C_6H_{12}O_6$, are present per dm^3 of solution A than of solution B.

(32) Illustrate the analogy between gases and dilute solutions by calculating the molar gas constant, R, in $atm\,dm^3\,K^{-1}\,mol^{-1}$ from the following data and comparing it with R for a perfect gas.

Cane sugar solutions at 15°C. Cane sugar $C_{12}H_{22}O_{11}$.

Concentration/g dm^{-3}	Osmotic pressure/mmHg
10	535
20	1016
40	2082
60	3075

(33) Given that the gas constant, R, has the value $0.082\,atm\,dm^3\,K^{-1}\,mol^{-1}$, and assuming the analogy between gases and dilute solutions, find, by using the formula $\Pi V = RT$, the osmotic pressure (in atmospheres) at 250°C of a solution containing 3.00 g of glycerol, $C_3H_8O_3$, in 100 cm^3 of solution.

(34) A solution containing 5.5 g of polystyrene in 1 dm^3 of benzene has an osmotic pressure of 10.5×10^{-4} atmospheres. Calculate a value for the mean relative molecular mass of the polystyrene chains. Polystyrene is poly(phenylethene).

6
Colligative Properties Involving Dissociation or Association

Theory

The treatment given in the preceding chapter assumes the absence of any dissociation of the solute into ions. The same basic ideas apply in the presence of dissociation but with the following modifications.

Suppose that a solute A ionizes in aqueous solution to produce n ions per molecule and that the degree of ionization is α at a given dilution and temperature. When ionic equilibrium has been established, an original mole of molecules of A will have produced $n\alpha$ moles of ions, leaving $(1 - \alpha)$ moles of covalent molecules. Since each of the ions is as effective in lowering of vapour pressure and related phenomena as a molecule, the total number of moles of effective particles is $(1 - \alpha + n\alpha)$ for each mole of the originally covalent molecules of A. Consequently the following relationships hold:

$$\frac{(1 - \alpha + n\alpha)}{1} = \frac{\text{Lowering of v.p. observed}}{\text{Lowering of v.p. calculated for no ionization}}$$

$$= \frac{\text{B.p. elevation observed}}{\text{B.p. elevation calculated for no ionization}}$$

$$= \frac{\text{F.p. depression observed}}{\text{F.p. depression calculated for no ionization}}$$

$$= \frac{\text{Osmotic pressure observed}}{\text{Osmotic pressure calculated for no ionization}}$$

When the solute is a strong electrolyte, α is the apparent degree of dissociation into ions.

When molecules associate into larger molecules or ions combine to form complex ions, there are fewer particles present than expected, therefore the observed effect is less than expected.

Examples

(1) *A molar solution of a weak, monobasic acid has a freezing point of $-1.93°C$ at atmospheric pressure, $101\,300\,N\,m^{-2}$. What is the degree of ionization of the acid and the osmotic pressure at $12°C$ of this molar solution? ($K = 1.86°C$ per $1000\,g$ of water. Assume that the molar volume of a gas is $22.4\,dm^3$ at s.t.p. and that the ionization of the acid is unchanged by the temperature rise to $12°C$.)*

If this molar solution contains covalent acid only, it must have a freezing point depression of $1.86°C$ (by definition of K). A monobasic acid gives two ions per molecule, $HA \rightleftharpoons H^+ + A^-$.

Consequently, if the degree of ionization of the acid is actually α,

$$\frac{(1-\alpha) + 2\alpha}{1} = \frac{1.93}{1.86}$$

i.e.,
$$(1 + \alpha) = 1.038$$

so
$$\alpha = 0.038$$

If the acid is covalent,

1 mole of the acid in $22.4\,dm^3$ of solution at $0°C$ gives an osmotic pressure of $101\,300\,N\,m^{-2}$

so

1 mole of the acid in $1\,dm^3$ of solution at $12°C$ gives an osmotic pressure of

$$101\,300 \times \frac{22.4}{1} \times \frac{285}{273} = 237\,000\,N\,m^{-2}$$

Since $\alpha = 0.038$, the actual osmotic pressure of the solution is

$$237\,000 \times (1 - \alpha + 2\alpha)$$

i.e.,
$$237\,000 \times 1.038 = 245\,900\,N\,m^{-2}$$

(2) *A solution containing $5.40\,g$ of strontium chloride, $SrCl_2$, in $170\,g$ of water freezes at $-0.982°C$ at $101\,300\,N\,m^{-2}$ pressure. Calculate the apparent degree of dissociation of the salt. ($K = 1.86°C$ per $1000\,g$ of water.)*

The mass of one mole of $SrCl_2 = 88 + 71 = 159\,g$. If the salt were undissociated,

$159\,g$ of $SrCl_2$ in $1000\,g$ of water depress the f.p. by $1.86°C$

so

$5.40\,g$ of $SrCl_2$ in $170\,g$ of water depress the f.p. by

$$1.86 \times \frac{5.40}{159} \times \frac{1000}{170} = 0.372°C$$

One mole of $SrCl_2$ will dissociate to produce three moles of ions:
$$Sr^{2+}(Cl^-)_2 \rightleftharpoons Sr^{2+} + 2Cl^-$$
If α is the apparent degree of dissociation,
$$\frac{(1-\alpha)+3\alpha}{1} = \frac{0.982}{0.372}$$
From this, $\qquad 1 + 2\alpha = 2.64$
i.e., $\qquad \alpha = 0.82$

i.e., the strontium chloride is apparently 82% dissociated.

(3) *By dissolving 8.5 g of common salt in 125 g of water at a certain temperature, the vapour pressure was depressed from $2666\,N\,m^{-2}$ to $2567\,N\,m^{-2}$. Calculate the degree of dissociation of common salt at this dilution and temperature.*

Let M be the apparent relative molecular mass of sodium chloride at the given dilution. Applying the vapour pressure formula quoted on p. 36,

$$\frac{2666 - 2567}{2666} = \frac{\dfrac{8.5}{M}}{\dfrac{8.5}{M} + \dfrac{125}{18}}$$

Solving this equation gives $M = 31.7$.

One mole of sodium chloride dissociates to produce two moles of ions:
$$Na^+Cl^- \rightleftharpoons Na^+ + Cl^-$$
If the apparent degree of dissociation of the salt is α,
$$\frac{(1-\alpha)+2\alpha}{1} = \frac{58.5}{31.7}$$
From this, $\qquad 1 + \alpha = 1.83$
i.e., $\qquad \alpha = 0.83$

That is, the salt is apparently 83% dissociated.

Problems on Colligative Properties Involving Dissociation and Association

(*Relative atomic masses will be found on page* 132)

(1) The vapour pressure of water at 20°C is $2333\,N\,m^{-2}$. An aqueous solution, containing 10.2 g of sodium nitrate in 120 g of water was found to have a vapour pressure of $2253\,N\,m^{-2}$ at 20°C. Calculate the apparent degree of dissociation of the sodium nitrate.

(2) The vapour pressure of water at 40°C is 4240 N m^{-2}. What would be the effect on this vapour pressure of dissolving 11.1 g of calcium chloride in 90 g of water? Assume that calcium chloride is 97% dissociated under these conditions.

(3) 1.00 g of potassium chloride was dissolved in 130 g of water, raising the boiling point by 0.10°C. Calculate the apparent degree of dissociation of the salt expressed as a percentage. ($K = 0.52$°C per 1000 g of water.)

(4) In a solution containing 1.70 g of silver nitrate in 160 g of water, the salt is apparently 83% dissociated. Calculate the freezing point of the solution. ($K = 1.86$°C per 1000 g of water.)

(5) A solution containing 1.20 g of strontium chloride in 12 g of water freezes at −0.327°C. Calculate the apparent degree of dissociation of strontium chloride at this dilution. ($K = 1.86$°C per 1000 g of water.)

(6) A 0.1M solution of dichloroethanoic acid freezes at −0.32°C. Calculate the degree of ionization of dichloroethanoic acid under these conditions. ($K = 1.86$°C per 1000 g of water.)

(7) 0.01M sodium chloride solution is apparently 92.5% dissociated at 18°C. Calculate the osmotic pressure of the solution at this temperature in N m^{-2}.

(8) A molar solution of potassium chloride has an osmotic pressure of 4 236 000 N m^{-2} at 20°C. Calculate the apparent degree of dissociation of potassium chloride at this dilution and temperature.

(9) A solution containing 3.00 g of calcium nitrate in 100 cm^3 of solution had an osmotic pressure of 1 134 000 N m^{-2} at 12°C. Calculate the apparent degree of dissociation of calcium nitrate at this dilution and temperature.

(10) A 0.01M solution of methanoic acid had an osmotic pressure of 28 350 N m^{-2} at 25°C. Calculate the degree of ionization of methanoic acid at this dilution and temperature.

(11) A certain calcium chloride solution has the same freezing point as a solution containing 3.00 g of sodium chloride in 1000 g of water. Calculate the mass of anhydrous calcium chloride per 100 g of water in the given solution, assuming that, at these dilutions, both salts are entirely dissociated.

(12) Calculate the apparent dissociation of calcium nitrate in an aqueous solution of the given dilution from the following data:

The freezing point of a solution of 4.60 g of cane sugar (relative molecular mass 342) in 134 g of water is −0.187°C.

The freezing point of a solution containing 0.18 moles of calcium nitrate in 1000 g of water is −0.840°C.

Both measurements made at 101 300 N m^{-2} pressure.

(13) A determination by Landsberger's method gave the following results:

Boiling point of pure water	99.66°C
Boiling point of an aqueous solution containing 0.75 g of urea (CON_2H_4)	99.92°C
Volume of aqueous solution containing the urea	25 cm^3
Boiling point of an aqueous solution containing 0.97 g of sodium chloride	100.15°C
Volume of solution containing the sodium chloride	34 cm^3

Pressure constant throughout.

Calculate, from these data alone, the apparent dissociation of the sodium chloride in the given solution.

(14) A 0.2M solution of sodium chloride freezes at −0.67°C. What would be the osmotic pressure of this solution at 8°C in N m^{-2}? Assume that the dissociation of the salt is unaffected by the temperature change. ($K = 1.86$°C per 1000 g of water.)

(15) What mass of glycerol would be necessary to produce the same anti-freezing effect in a dm^3 of water as 20 g of sodium chloride? Assume the sodium chloride is 97% dissociated.

(16) 0.47 g of a non-electrolyte, of relative molecular mass 58, in 90 g of water had a freezing point of −0.167°C. A solution of zinc chloride containing 2.90 g of the anhydrous salt in 1000 g of water had a freezing point of −0.111°C. Calculate the apparent dissociation of zinc chloride in this solution, using the above data alone.

(17) 0.31 g of methanol, CH_3OH, depresses the freezing point of 80 g of benzene by 0.30°C. What is the molecular state of methanol in benzene? ($K = 5.0$°C per 1000 g of benzene.)

(18) 0.75 g of ethanoic acid, CH_3COOH, when dissolved in 125 g of benzene, depresses the freezing point of benzene by 0.255°C. What is the molecular state of ethanoic acid in benzene solution? (K = 5.0°C per 1000 g of benzene.)

(19) Account for the following observations: A solution of iron(III) potassium sulphate, containing 0.1 mole of $KFe(SO_4)_2$ in 1000 g of water, and a solution of potassium hexacyanoferrate(III), containing 0.1 mole of $K_3Fe(CN)_6$ in 1000 g of water, both freeze at −0.744°C. ($K = 1.86$°C per 1000 g of water.)

(20) Assuming potassium iodide to be completely dissociated under the given conditions, at what temperature would you expect a solution, containing 1.66 g of potassium iodide in 100 g of water, to freeze? Account for the fact that when 2.27 g of mercury(II) iodide is added to the above solution and the mixture stirred until all the mercury(II) iodide dissolves, the solution now freezes at −0.27°C. ($K = 1.8$°C per 1000 g of water.)

7

Relative Atomic and Molecular Masses by Relative Vapour Density and Other Methods

Theory

The relative vapour density of a gas or vapour is defined in the following way:

$$\frac{\text{Relative vapour density}}{\text{of a gas or vapour}} = \frac{\text{Mass of 1 volume of gas or vapour}}{\text{Mass of 1 volume of hydrogen}}$$

(both volumes being measured at the same temperature and pressure)

The relative molecular mass of a gas is related to its relative vapour density by the expression:

$$\text{Relative molecular mass} = 2 \times \text{relative vapour density}$$

The molar volume of any gas is $22.4\,dm^3$ at s.t.p., i.e., the number of grams of the gas which occupy $22.4\,dm^3$ at s.t.p. is the relative molecular mass of the gas.

Examples

The following examples and problems involve the comparison of pressures and therefore, if preferred, the unit mmHg could be used rather than the SI unit, $N\,m^{-2}$ or Pa.

(1) METHOD OF V. MEYER

$0.143\,g$ *of a volatile liquid, when vaporized in a V. Meyer apparatus, displaced* $30.4\,cm^3$ *of air at* $17°C$ *and* $99\,350\,N\,m^{-2}$ *($745\,mmHg$) pressure. Calculate the relative vapour density and relative molecular mass of the liquid. The vapour pressure of water at* $17°C$ *is* $2000\,N\,m^{-2}$ *($15\,mmHg$).*

Method (*a*), using the fact that $1\,dm^3$ of hydrogen at s.t.p. has a mass of $0.09\,g$.

Volume of air displaced at s.t.p.

$$= 30.4 \times \frac{273}{290} \times \frac{(99\,350 - 2000)}{101\,300} = 27.5 \,\text{cm}^3$$

Mass of this volume of hydrogen at s.t.p.

$$= \frac{0.09}{1000} \times 27.5 = 0.00248 \,\text{g}$$

Relative vapour density of the substance

$$= \frac{0.143}{0.00248} = 57.7$$

Relative molecular mass of the liquid

$$= 57.7 \times 2 = 115.4$$

Method (b), using the fact that the molar volume of a gas is $22.4\,\text{dm}^3$ at s.t.p. As in method (a) the volume of displaced air converted to s.t.p. is $27.5\,\text{cm}^3$. The mass of substance which displaced this air was $0.143\,\text{g}$. Therefore,

mass of substance needed to displace $22.4\,\text{dm}^3$ of air at s.t.p.

$$= 0.143 \times \frac{22\,400}{27.5} = 116.5 \,\text{g}$$

That is, the relative molecular mass of the liquid is 116.5 and its relative vapour density is half this figure, i.e., 58.3.

(Notice that the two calculations give results differing by about 1%. The reason is that the datum, $1\,\text{dm}^3$ of hydrogen at s.t.p. has a mass of $0.09\,\text{g}$, is used; this is equivalent to employing the standard, $H = 1.000$. If molar volume $= 22.4\,\text{dm}^3$ at s.t.p. is taken, this is equivalent to using $H = 1.008$. The second of these is greater by nearly 1%.)

(2) METHOD OF DUMAS

Calculate the relative vapour density and relative molecular mass of a volatile liquid X from the following data:

Mass of Dumas bulb full of air at $10°C$ and $102\,700\,N\,m^{-2}$ ($770\,mmHg$)	$= 25.700\,\text{g}$
Mass of bulb sealed full of vapour of X at $100°C$ and $102\,700\,N\,m^{-2}$ ($770\,mmHg$)	$= 25.759\,\text{g}$
Mass of bulb full of water	$= 265.1\,\text{g}$

Given: $1\,\text{dm}^3$ of hydrogen at s.t.p. has a mass of $0.09\,\text{g}$

$1\,\text{dm}^3$ of air at s.t.p. has a mass of $1.293\,\text{g}$

(a) Find the capacity of the bulb from the mass of water contained.

Mass of water in the bulb
$$= (265.1 - 25.7) = 239.4\,g$$
That is, capacity of bulb $= 239.4\,cm^3$

(b) Convert the volume of contained air to s.t.p., find its mass and, hence, the upthrust of air upon the bulb.

Volume of air in bulb at s.t.p.
$$= 239.4 \times \frac{273}{283} \times \frac{102\,700}{101\,300}\,cm^3$$

Mass of air in the bulb
$$= 239.4 \times \frac{273}{283} \times \frac{102\,700}{101\,300} \times \frac{1.293}{1000} = 0.303\,g$$
this being the upthrust of air upon the bulb.

(c) Find the mass of vapour sealed into the bulb.

Mass of vapour in bulb
$$= (25.759 - 25.700 + 0.303) = 0.362\,g$$

(d) Convert the volume of vapour to s.t.p.

Volume of vapour at s.t.p.
$$= 239.4 \times \frac{273}{373} \times \frac{102\,700}{101\,300} = 177.5\,cm^3$$

(e) Calculate the relative vapour density and relative molecular mass of X.

Mass of volume of hydrogen equal to volume of vapour at s.t.p.
$$= 177.5 \times \frac{0.09}{1000} = 0.0160\,g$$

Relative vapour density of X
$$= \frac{0.362}{0.0160} = 22.6$$

Relative molecular mass of X
$$= 22.6 \times 2 = 45.2$$

If use of molar volume $= 22.4\,dm^3$ at s.t.p. is preferred, the last few lines of the calculation will run:

$177.5\,cm^3$ of vapour of X at s.t.p. have a mass of $0.362\,g$, so

22 400 cm^3 of vapour of X at s.t.p. have a mass of

$$0.362 \times \frac{22\,400}{177.5} = 45.7 \text{ g}$$

Consequently, the relative molecular mass of X is 45.7 and the relative vapour density is half of this, that is, 22.9. For an explanation of the higher value, see the V. Meyer calculation above.

(3) CANNIZZARO'S METHOD

If a large number of volatile compounds of an element are selected for analysis, then it is likely that at least one of the compounds will contain one atom of the element per molecule. Therefore, the relative atomic mass of the element is probably the least mass of the element in the relative molecular mass of one of its compounds.

A number of gaseous or volatile silicon compounds yield the following data. Calculate the probable relative atomic mass of silicon.

Compound	Percentage of silicon by mass	Relative vapour density of compound
A	87.5	16
B	90.3	31
C	96.6	29
D	91.3	46
E	16.5	85

The relative molecular mass of a compound being twice its relative vapour density, the following data can be calculated:

Compound	Relative molecular mass	Mass of silicon in the relative molecular mass
A	$2 \times 16 = 32$	$\frac{87.5}{100} \times 32 = 28.0$
B	$2 \times 31 = 62$	$\frac{90.3}{100} \times 62 = 56.0$
C	$2 \times 29 = 58$	$\frac{96.6}{100} \times 58 = 56.0$
D	$2 \times 46 = 92$	$\frac{91.3}{100} \times 92 = 84.0$
E	$2 \times 85 = 170$	$\frac{16.5}{100} \times 170 = 28.0$

All the figures in the final column are 28 or multiples of 28. The probable relative atomic mass of silicon is, therefore, 28. This is the true relative atomic mass if at least one of the compounds considered has only *one* atom of silicon in its molecule.

(4) ISOMORPHISM

A metal, X, *forms an alum which includes the element potassium and is isomorphous with ordinary alum. If ordinary alum is known to be* $KAl(SO_4)_2 \cdot 12H_2O$ *and the alum of* X *contains 10.42% of* X, *calculate the relative atomic mass of* X.

Being isomorphous with ordinary alum, the alum of X will be expected to have an analogous structure and to be $KX(SO_4)_2 \cdot 12H_2O$. The formula mass of this alum is $(447 + A)$, where A is the relative atomic mass of X. Therefore,

$$\frac{A}{447 + A} \times 100 = 10.42$$

From this, $\qquad 100A = 10.42(447 + A)$

i.e., $\qquad 100A = 4658 + 10.42A$

and $\qquad A = \dfrac{4658}{89.58} = 52.0$

(5) DIFFUSION

Graham's Law states: The rates of diffusion of gases under the same conditions are inversely proportional to the square roots of their densities.

A certain volume of oxygen diffused from a given apparatus in 120 seconds. In the same conditions, the same volume of a gas, X, *diffused in 112 seconds. Calculate the relative molecular mass of* X.

If the volume of the two gases diffusing is $v\,\text{cm}^3$, the rates of diffusion are:

$$\text{Oxygen } \frac{v}{120} \text{ cm}^3\text{s}^{-1} \qquad \text{Gas } X \, \frac{v}{112} \text{ cm}^3\text{s}^{-1}$$

The densities of gases are directly proportional to their relative molecular masses. The relative molecular mass of oxygen, O_2, is 32. If the relative molecular mass of X is M, Graham's Law requires the relation:

$$\frac{\text{Rate of diffusion of oxygen}}{\text{Rate of diffusion of } X} = \frac{\dfrac{v}{120}}{\dfrac{v}{112}} = \frac{\sqrt{M}}{\sqrt{32}}$$

Therefore,

$$\frac{M}{32} = \frac{112^2}{120^2} \quad \text{from which} \quad M = \frac{32 \times 112^2}{120^2} = 27.9$$

The gas X may be nitrogen or carbon monoxide or ethene.

Problems on Relative Atomic and Molecular Masses by Relative Vapour Density and Other Methods

(1) Calculate the relative molecular mass of a gas X from the following data:

Mass of evacuated globe	11.270 g
Mass of globe filled with hydrogen	11.630 g
Mass of globe filled with gas X	19.180 g

Temperature and pressure were constant throughout the experiment.

(2) A student's determination of the relative molecular mass of oxygen yielded the following data:

Mass of flask and potassium chlorate before heating	12.500 g
Mass of flask and potassium chloride after heating	12.134 g
Volume of oxygen evolved	276 cm^3
Temperature of gas	17°C
Pressure of gas	100 000 N m^{-2} (750 mmHg)

Calculate the relative molecular mass of oxygen.

(3) Calculate the relative vapour density and hence find the relative molecular mass of a volatile organic liquid, from the following data, obtained using a Dumas bulb.

Mass of bulb and air	30.78 g
Mass of bulb and vapour after sealing	31.03 g
Mass of bulb full of water (and piece broken off)	209.0 g
Temperature of laboratory	20°C
Temperature of the bath at moment of sealing	73°C
Pressure	100 700 N m^{-2} (755 mmHg)

(4) Calculate the relative molecular mass of carbon dioxide from the following data:

Mass of evacuated globe	27.30 g
Mass of globe full of carbon dioxide at 17°C and 101 100 N m^{-2} (758 mmHg)	28.40 g
Mass of globe full of water	621.40 g

(5) Find the relative molecular mass of a volatile liquid from the following Victor Meyer experiment:

Mass of small stoppered tube	0.672 g
Mass of small stoppered tube and liquid	0.783 g
Volume of air displaced	35.8 cm^3
Temperature	12°C
Pressure	100 800 N m^{-2} (756 mmHg)
Vapour pressure of water vapour at 12°C	1466 N m^{-2} (11 mmHg)

(6) Find the relative vapour density and the relative molecular mass of trichloromethane (chloroform) from the following Dumas experiment:

Mass of globe filled with air at 17°C and 100 700 N m^{-2} (755 mmHg)	25.230 g
Mass of globe filled with vapour at 80°C and 100 700 N m^{-2} (755 mmHg)	25.697 g
Mass of globe filled with water	184.73 g

(7) 0.119 g of a volatile liquid when vaporized in a Victor Meyer's apparatus displaced 25.4 cm^3 of air at 17°C and 99 350 N m^{-2} (745 mmHg). Calculate the relative molecular mass of the liquid.
Vapour pressure of water at 17°C is 2000 N m^{-2} (15 mmHg).

(8) A small volume of a volatile liquid was injected, by means of a hypodermic syringe, into a large graduated syringe surrounded by a steam jacket. From the following results, calculate the relative molecular mass of the liquid:

Mass of hypodermic syringe before injection	17.436 g
Mass of hypodermic syringe after injection	17.025 g
Volume of vapour produced in large heated syringe	70.2 cm^3
Temperature of steam bath	98°C
Atmospheric pressure	100 000 N m^{-2} (750 mmHg)

(9) The relative vapour density of a volatile liquid was determined by injecting a small quantity of the liquid into a large heated syringe. The following results were recorded:

Mass of hypodermic syringe before injection	17.216 g
Mass of hypodermic syringe after injection	17.100 g
Volume of vapour formed	45.4 cm^3

Temperature of heated syringe . . . 96°C
Pressure . . . . 101 600 N m^{-2}
(762 mmHg)

Calculate the relative vapour density and relative molecular mass of the liquid.

(10) 140 cm^3 of water vapour measured at 150°C and 100 700 N m^{-2} (755 mmHg) pressure have a mass of 0.071 g. Show that these figures are in accordance with the formula H_2O for steam.

(11) A certain element X forms a large number of gaseous and volatile compounds of which the following are representative:

Compound with	% by mass of X	Relative vapour density of compound
Carbon	57.1	14
Carbon	72.7	22
Sulphur	50.0	32
Sulphur	60.0	40
Hydrogen	88.9	9
Nitrogen	53.3	15

What is the probable relative atomic mass of X?

(12) A certain element M forms a large number of gaseous and volatile compounds of which the following are a sample:

Compound with	% of M by mass	R.V.D. of compound
Hydrogen	75.0	8
Hydrogen	92.3	39
Hydrogen	81.8	22
Oxygen	42.9	14
Oxygen	27.3	22
Hydrogen and chlorine	10.0	59.8
Hydrogen and oxygen	52.2	23

What is the probable relative atomic mass of M?

(13) Give with reasons the value you would assign to the relative atomic mass of sulphur from the following analyses of its volatile compounds:

	R.V.D.	% by mass of sulphur
Carbon disulphide	38	84
Sulphur dioxide	32	50
Hydrogen sulphide	17	94
Sulphur(VI) oxide	40	40

(14) 0.56 g of a metal when treated with dilute acid liberated 560 cm^3 of hydrogen measured at 15°C and 100 000 N m^{-2} (750 mmHg). The specific heat capacity of the metal is $1.05\,\text{J}\,\text{g}^{-1}\,\text{K}^{-1}$. What is the relative atomic mass of the metal?

(15) The following are analyses of two oxides of a metal:

	I	II
Mass of oxide	2.92 g	4.33 g
Mass of metal	2.71 g	3.75 g

The specific heat capacity of the metal is $0.131\,\text{J}\,\text{g}^{-1}\,\text{K}^{-1}$. Calculate the mean relative atomic mass as suggested by these results and write formulae for the compounds which were analysed.

(16) Beryllium chloride was known to contain 11.3% of beryllium but the oxidation number of the metal was uncertain. The chloride was found to have a relative vapour density of 40. Determine the oxidation number of beryllium in this compound and hence its relative atomic mass.

(17) A specimen of Dialogite (crystalline manganese(II) carbonate) gave the following analysis:

Mn 47.83%. C 10.43%. O 41.74% by mass.

The crystals were isomorphous with calcite (calcium carbonate). Calculate the relative atomic mass of manganese.

(18) Potassium sulphate and potassium selenate are isomorphous. Calculate the relative atomic mass of selenium from the following analyses:

Potassium sulphate, K = 44.83%, S = 18.40%, O = 36.77% by mass.

Potassium selenate, K = 35.30%, Se = 35.75%, O = 28.96% by mass.

(19) In a diffusion experiment, the time required for water to cover the space between two marks on a tube as oxygen diffused out of it was 240 seconds. In identical conditions, the time required for another gas, X, was three and three-quarters minutes. Calculate the relative molecular mass of X.

(20) A certain volume of hydrogen diffuses from an apparatus in one minute. Calculate the time required for the diffusion of the same volume of ozonized oxygen, containing 10% of ozone by volume, from the apparatus under identical conditions. (Ozone is O_3, trioxygen.)

8

Thermal Dissociation

Theory

Thermal dissociation is the breaking up of molecules of an element or compound by heat into products simpler than themselves, such that, on cooling, the products recombine completely. Examples are:

$$N_2O_4 \rightleftharpoons 2NO_2; \qquad 2HI \rightleftharpoons H_2 + I_2$$
$$NH_4Cl \rightleftharpoons NH_3 + HCl; \qquad PCl_5 \rightleftharpoons PCl_3 + Cl_2$$
$$I_2 \rightleftharpoons I + I$$

The total *mass* of material in any given case remains *constant* throughout; the increase in the number of molecules causes, at constant pressure, increase of volume. Consequently, thermal dissociation is accompanied by a decrease in the relative vapour density of the material. From this fact, the degree of dissociation can usually be calculated.

Examples

(1) *The relative vapour density of nitrogen dioxide at 105°C is 24.0. Calculate the degree of dissociation (by mass) of the gas in these conditions.*

The dissociation equilibrium is:

$$N_2O_4 \rightleftharpoons 2NO_2$$

Originally   1 mole   —
At equilibrium $(1 - \alpha)$ mole 2α mole,
where α is the degree of dissociation.

$M_r(N_2O_4) = 92$, i.e., the relative density of pure N_2O_4 is $\frac{92}{2} = 46$

The volumes of the gases, at constant temperature, are directly proportional to the numbers of moles present, but the mass remains constant. Consequently,

$$\frac{(1 - \alpha) + 2\alpha}{1} = \frac{46}{24.0}$$

From this, $\qquad 1 + \alpha = 1.92$

and $\qquad \alpha = 0.92$

That is, in the conditions stated, the gas is 92% dissociated.

(2) *If phosphorus pentachloride is 20% dissociated at a certain temperature, calculate the relative vapour density of the equilibrium mixture of the pentachloride, trichloride and chlorine at this temperature.*

$$PCl_5 \rightleftharpoons PCl_3 + Cl_2$$

Originally	1 mole	—	—
At equilibrium	$(1 - \alpha)$	α	α mole

where α is the degree of dissociation.

As the mass of material remains constant and the volume is proportional, at constant temperature and pressure, to the number of moles present,

$$\frac{(1 - \alpha) + 2\alpha}{1} = \frac{\frac{1}{2}(208.5)}{x}$$

where x is the required relative vapour density.

That is, $\qquad (1 + \alpha) = \dfrac{104.25}{x}$

where α is 20% or 0.2. This gives

$$(1 + 0.2)x = 104.25$$

or $\qquad x = \dfrac{104.25}{1.2} = 95.2$

That is, the relative vapour density of the dissociated mixture is 95.2.

(3) *The relative vapour density of iodine at 1 atmosphere pressure and 1250°C is 87. Calculate the percentage dissociation of iodine, I_2, into atoms in these conditions.*

$$I_2 \rightleftharpoons I + I$$

Originally	1 mole	—	—
At equilibrium	$(1 - \alpha)$ mole	α	α mole

where the degree of dissociation is α.

As the mass of iodine remains constant and the volume is directly proportional to the number of moles at constant temperature and pressure,

$$\frac{(1-\alpha)+2\alpha}{1}=\frac{\frac{1}{2}(254)}{87}=\frac{127}{87}$$

From this, $\qquad 1+\alpha=1.46$

and $\qquad\qquad\quad \alpha=0.46$

That is, the iodine is 46% dissociated.

Problems on Thermal Dissociation

(Relative atomic masses will be found on page 132)

(1) At 400°C the relative vapour density of ammonium chloride is 14.5. Calculate the degree of dissociation of ammonium chloride into ammonia and hydrogen chloride at this temperature.

(2) 1 g of hydrogen iodide was heated to a certain temperature and the products suddenly cooled. The iodine which had been liberated required 15 cm^3 of 0.1M sodium thiosulphate for complete action. Calculate the percentage by mass of the hydrogen iodide which remained undissociated at that temperature.

(3) 2.5 g of calcium carbonate are placed in a globe of capacity 850 cm^3. The globe is evacuated and heated to a temperature of 800°C at which temperature the carbon dioxide evolved exerts a pressure of 26 660 N m^{-2} (200 mmHg). Calculate the percentage of the calcium carbonate which is decomposed.

(4) What will be the relative vapour density of phosphorus pentachloride at 220°C if, at that temperature, 66% by mass is decomposed into phosphorus trichloride and chlorine?

(5) What will be the percentage gravimetric composition of the mixture obtained from hydrogen iodide which has been heated to 450°C? At this temperature 23% by mass of the gases consists of molecules of hydrogen and iodine.

(6) When iodine is heated to a high temperature, it dissociates into atoms. Draw a graph from the following data showing the relation between temperature and percentage of iodine dissociated.

Temp.	1100°C	1200°C	1300°C	1400°C	1500°C
R.V.D.	97.5	90	83.4	78	73

From your graph estimate the temperature at which the iodine is 80% dissociated.

(7) 1 cm^3 of water measured at 4°C is heated until it has attained a temperature of 2500°C. If steam at this temperature is 3.98% by mass dissociated into molecules of hydrogen and oxygen, calculate the volume the gases should occupy. Assume that the gas laws apply at the temperature given. Pressure 101 300 N m^{-2} (760 mmHg) throughout.

(8) What volume of 0.165M sodium thiosulphate would be required to react with the iodine liberated from 3 g of hydrogen iodide if the latter were heated to a temperature where it is 16% dissociated? What mass of hydrogen would remain uncombined?

(9) At 27°C dinitrogen tetraoxide is 20% dissociated into nitrogen dioxide. 1 dm^3 of this mixture was heated to (a) 100°C and (b) 600°C. At 100°C the dinitrogen tetraoxide is 90% dissociated and at 600°C it is completely dissociated into nitrogen oxide and oxygen. Calculate the volume the gases would occupy under conditions (a) and (b). All measurements made at 101 300 N m^{-2} (760 mmHg).

(10) At a certain temperature and under identical conditions, the volumes of oxygen and dinitrogen tetraoxide diffusing from an apparatus in a given time were in the ratio of 3 to 2. Calculate the degree of dissociation of dinitrogen tetraoxide at this temperature.

$$N_2O_4 \rightleftharpoons 2NO_2$$

(11) The relative vapour density of a sample of ozone at a certain temperature was 18.3. What is the degree of dissociation of ozone into oxygen under these conditions? (Ozone is O_3, trioxygen.)

9

Energetics

Theory

SIGN CONVENTION AND UNITS

During an exothermic reaction, heat is given out to the surroundings and the enthalpy change (heat of reaction at constant pressure), ΔH, is said to be negative as the system has lost energy. During an endothermic reaction ΔH is positive as the system has gained energy.

Enthalpy changes are measured in kilojoules per mole ($kJ\,mol^{-1}$).

DEFINITIONS

The standard enthalpy change for a reaction, $\Delta H^\ominus$ is equal to the heat evolved or absorbed when the molar quantities, represented by the equation, react at a pressure of $101\,300\,N\,m^{-2}$ ($760\,mmHg$) and a temperature of $298\,K$, the substances being in their normal physical state for these conditions.

The standard enthalpy of formation, $\Delta H_f^\ominus$, is the heat change which occurs when 1 mole of a compound is formed from its elements under the standard conditions of temperature and pressure ($101\,300\,N\,m^{-2}$ and $298\,K$).

The standard enthalpy of combustion, $\Delta H_c^\ominus$, is the heat evolved when 1 mole of an element or compound is completely burned in oxygen under standard conditions.

The standard enthalpy of atomization is the heat required to convert an element, in its normal state and under standard conditions, into 1 mole of gaseous atoms.

The standard enthalpy of hydrogenation is the heat change which occurs when 1 mole of an unsaturated compound is converted to the corresponding saturated compound by reaction with hydrogen under standard conditions.

The standard enthalpy of neutralization is the heat evolved when a dilute solution containing sufficient mass of acid to provide 1 mole of

H^+ ions reacts with a dilute solution containing sufficient alkali to provide 1 mole of OH^- ions, under standard conditions.

The lattice energy of an ionic crystal is the standard enthalpy of formation of 1 mole of the crystal from its ions in the gas phase.

HESS'S LAW

The total energy change resulting from a chemical reaction is dependent only on the initial and final states of the reactants and is independent of the reaction route.

Examples

(1) *Calculate the standard enthalpy of formation of propane, C_3H_8, from the data:*

$C(graphite) + O_2(g) \rightarrow CO_2(g) \quad \Delta H^\ominus = -393.5 \text{ kJ mol}^{-1}$

$H_2(g) + \tfrac{1}{2}O_2(g) \quad \rightarrow H_2O(l) \quad \Delta H^\ominus = -285.9 \text{ kJ mol}^{-1}$

The standard enthalpy of combustion of propane is $-2220 \text{ kJ mol}^{-1}$.

The equation for the combustion of propane is:

$C_3H_8(g) + 5O_2(g) \rightarrow 3CO_2(g) + 4H_2O(l) \quad \Delta H^\ominus = -2220 \text{ kJ mol}^{-1}$

An energy cycle can be written as follows:

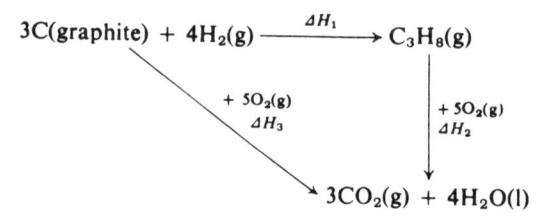

$$3C(graphite) + 4H_2(g) \xrightarrow{\Delta H_1} C_3H_8(g)$$

$$\searrow \substack{+ 5O_2(g) \\ \Delta H_3} \qquad \downarrow \substack{+ 5O_2(g) \\ \Delta H_2}$$

$$3CO_2(g) + 4H_2O(l)$$

According to Hess's Law the total enthalpy change during the formation of the combustion products by one route is equal to the total enthalpy change when they are formed by another route, i.e.,

$$\Delta H_1 + \Delta H_2 = \Delta H_3 \qquad \qquad (i)$$

$\Delta H_1 = \Delta H_f^\ominus$, the standard enthalpy of formation of propane,

$\Delta H_2 =$ the standard enthalpy of combustion of propane,

$\Delta H_3 = 3 \times \Delta H_f^\ominus$ of $CO_2(g) + 4 \times \Delta H_f^\ominus$ of $H_2O(l)$.

Substituting in equation (i),

$$\Delta H_f^\ominus \text{ of } C_3H_8 - 2220 = 3(-393.5) + 4(-285.9)$$

$$\Delta H_f^\ominus = -104.1 \text{ kJ mol}^{-1}$$

(2) *The standard enthalpy of formation of phenol is $-209.3\,kJ\,mol^{-1}$. Calculate its standard enthalpy of combustion, given:*

$$C(graphite) + O_2(g) \rightarrow CO_2(g) \quad \Delta H^\ominus = -393.5\,kJ\,mol^{-1}$$
$$H_2(g) + \tfrac{1}{2}O_2(g) \rightarrow H_2O(l) \quad \Delta H^\ominus = -285.9\,kJ\,mol^{-1}$$

The equation for the combustion of phenol is:

$$C_6H_5OH(s) + 7O_2(g) \rightarrow 6CO_2(g) + 3H_2O(l)$$

An energy cycle can be written as follows:

$$6C(graphite) + 3H_2(g) + \tfrac{1}{2}O_2(g) \xrightarrow{\Delta H_1} C_6H_5OH(s)$$

with $+7O_2(g)$, ΔH_3 on the left branch and $+7O_2(g)$, ΔH_2 on the right branch, both leading to:

$$6CO_2(g) + 3H_2O(l)$$

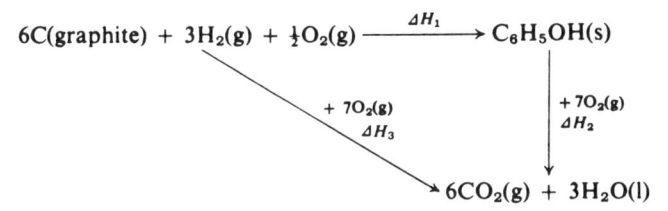

By Hess's Law,

$$\Delta H_1 + \Delta H_2 = \Delta H_3$$

ΔH_1 = the standard enthalpy of formation of phenol,
ΔH_2 = the standard enthalpy of combustion of phenol, $\Delta H_c^\ominus$,
ΔH_3 = $6 \times \Delta H_f^\ominus$ of $CO_2(g)$ + $3 \times \Delta H_f^\ominus$ of $H_2O(l)$.

Substituting,

$$-209.3 + \Delta H_c^\ominus = 6(-393.5) + 3(-285.9)$$
$$\Delta H_c^\ominus = -3009\,kJ\,mol^{-1}$$

(3) *Calculate the standard enthalpy change for the reaction:*

$$Al_2O_3(s) + 6Na(s) \rightarrow 2Al(s) + 3Na_2O(s)$$

given:

$$2Al(s) + 1\tfrac{1}{2}O_2(g) \rightarrow Al_2O_3(s) \quad \Delta H^\ominus = -1590\,kJ\,mol^{-1}$$
$$2Na(s) + \tfrac{1}{2}O_2(g) \rightarrow Na_2O(s) \quad \Delta H^\ominus = -422.6\,kJ\,mol^{-1}$$

An energy cycle can be written as follows:

$$Al_2O_3(s) + 6Na(s) \xrightarrow{\Delta H_1} 2Al(s) + 3Na_2O(s)$$

with ΔH_3 on the left (upward) and ΔH_2 on the right, both connecting to:

$$2Al(s) + 1\tfrac{1}{2}O_2(g) + 6Na(s)$$

From Hess's Law,

$$\Delta H_1 + \Delta H_3 = \Delta H_2$$

ΔH_1 = the standard enthalpy change for the reaction,
ΔH_3 = the standard enthalpy of formation of $Al_2O_3(s)$,
$\Delta H_2 = 3 \times$ the standard enthalpy of formation of $Na_2O(s)$.

Substituting,

$$\Delta H_1 - 1590 = 3(-422.6)$$
$$\Delta H_1 = 322.2 \text{ kJ mol}^{-1}$$

(4) *125 cm³ of 0.5M potassium hydroxide were mixed in a suitably lagged calorimeter, of mass 53.5 g, with 125 cm³ of 0.5M hydrochloric acid, both being at 14.00°C. The temperature (after suitable correction) reached a maximum of 17.30°C. Assuming the dilute solutions to have the same specific heat capacity as water ($4.18 \text{ J g}^{-1} {}^\circ C^{-1}$) and the calorimeter to have a specific heat capacity of $0.390 \text{ J g}^{-1} {}^\circ C^{-1}$. Calculate the enthalpy of neutralization.*

Heat capacity of calorimeter = $53.5 \times 0.390 \text{ J} {}^\circ C^{-1}$
Heat capacity of solution = $250 \times 4.18 \text{ J} {}^\circ C^{-1}$
Total heat capacity = $20.86 + 1045 \text{ J} {}^\circ C^{-1}$
= $1066 \text{ J} {}^\circ C^{-1}$
Total enthalpy change = $1066 \times 3.30 \text{ J}$
= 3518 J

125 cm³ of 0.5M KOH contains $0.5 \times \dfrac{125}{1000}$ moles of OH^-

125 cm³ of 0.5M HCl contains $0.5 \times \dfrac{125}{1000}$ moles of H^+

Therefore, the total enthalpy change for 1 mole of H^+ reacting with 1 mole of OH^-

$$= 3518 \times \frac{1}{0.5} \times \frac{1000}{125}$$
$$= 56\,290 \text{ J}$$

i.e., the enthalpy of neutralization is $56.29 \text{ kJ mol}^{-1}$.

(5) *The standard enthalpy of formation of methane is $-74.8 \text{ kJ mol}^{-1}$. The enthalpies of atomization of graphite and hydrogen are $+715$ and $+218 \text{ kJ mol}^{-1}$ of atoms in the gas phase. Calculate the mean $C-H$ bond energy.*

The following energy cycle can be written:

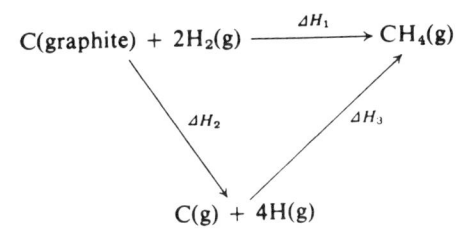

$$\text{C(graphite)} + 2\text{H}_2(g) \xrightarrow{\Delta H_1} \text{CH}_4(g)$$

with ΔH_2 going down from C(graphite) + 2H$_2$(g) to C(g) + 4H(g), and ΔH_3 going up from C(g) + 4H(g) to CH$_4$(g).

$$\text{C(g)} + 4\text{H(g)}$$

By Hess's Law,

$$\Delta H_1 = \Delta H_2 + \Delta H_3$$

ΔH_1 = the standard enthalpy of formation of methane,

ΔH_2 = the enthalpy of atomization of graphite + 4 × the enthalpy of atomization of hydrogen,

ΔH_3 = the enthalpy of formation of 1 mole of methane from gaseous atoms of carbon and hydrogen.

Substituting,

$$-74.8 = 715 + (4 \times 218) + \Delta H_3$$

$$\Delta H_3 = -1661.8 \text{ kJ mol}^{-1}$$

This is the energy evolved during the formation of four moles of C—H bonds from gaseous atoms of carbon and hydrogen. Therefore, the mean C—H bond energy

$$= -\frac{1661.8}{4} = -415.4 \text{ kJ mol}^{-1}$$

(Bond energies are derived here as the exothermic changes accompanying the formation of bonds but are usually tabulated and considered as the endothermic changes accompanying the breaking of bonds.)

(6) *Assuming the mean C—H bond energy to be* -415 kJ mol^{-1} *and the standard enthalpy of formation of ethane from gaseous atoms of carbon and hydrogen to be* $-2823 \text{ kJ mol}^{-1}$, *calculate the mean C—C bond energy per mole.*

The equation representing the formation of ethane from gaseous atoms is:

$$2\text{C(g)} + 6\text{H(g)} \rightarrow \text{C}_2\text{H}_6(g) \quad \Delta H^\ominus = -2823 \text{ kJ mol}^{-1}$$

This is the energy evolved during the formation of 1 mole of C—C bonds and 6 moles of C—H bonds, i.e.,

$$-2823 = -1(\text{C}-\text{C bond energy}) + (6 \times -415)$$

mean C—C bond energy = 333 kJ mol^{-1}.

(7) *Assuming the following mean bond energies,*

$$C\equiv C \quad 813\,kJ\,mol^{-1}$$
$$C-C \quad 346\,kJ\,mol^{-1}$$
$$C-H \quad 413\,kJ\,mol^{-1}$$
$$H-H \quad 436\,kJ\,mol^{-1}$$

predict the standard enthalpy of hydrogenation of ethyne (acetylene).

The equation for the hydrogenation is:

$$H-C\equiv C-H \;+\; 2(H-H) \;\rightarrow\; \begin{array}{c} H\;\;H \\ |\;\;\;| \\ H-C-C-H \\ |\;\;\;| \\ H\;\;H \end{array}$$

This change could be considered to take place by the breaking of 1 mole of $C\equiv C$ bonds and 2 moles of $H-H$ bonds which would require:

$$813 + (2 \times 436) = 1685\,kJ$$

This is then followed by the forming of 1 mole of $C-C$ bonds and 4 moles of $C-H$ bonds. This would result in

$$346 + (4 \times 413) = 1998\,kJ$$

being evolved. Therefore the enthalpy of hydrogenation of ethyne,

$$= 1685 - 1998 = -313\,kJ\,mol^{-1}$$

(8) *Calculate the lattice energy of sodium chloride from the following data: The standard enthalpy of formation of sodium chloride is $-411\,kJ\,mol^{-1}$, the standard enthalpies of atomization of sodium and chlorine are respectively $+108$ and $+121\,kJ\,mol^{-1}$ of gaseous atoms, the ionization energy of sodium is $+493\,kJ\,mol^{-1}$ and the electron affinity of chlorine is $-364\,kJ\,mol^{-1}$.*

The following Born-Haber energy cycle can be written:

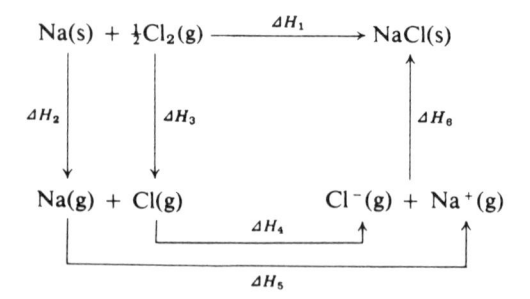

$$\begin{array}{ccc}
Na(s) + \tfrac{1}{2}Cl_2(g) & \xrightarrow{\Delta H_1} & NaCl(s) \\
\Delta H_2 \downarrow \;\; \Delta H_3 \downarrow & & \uparrow \Delta H_6 \\
Na(g) + Cl(g) & & Cl^-(g) + Na^+(g) \\
& \xrightarrow{\Delta H_4} & \uparrow \\
& \xrightarrow{\Delta H_5} &
\end{array}$$

By Hess's Law,

$$\Delta H_1 = \Delta H_2 + \Delta H_3 + \Delta H_4 + \Delta H_5 + \Delta H_6$$

ΔH_1 = standard enthalpy of formation of
sodium chloride = $-411\,\text{kJ mol}^{-1}$

ΔH_2 = standard enthalpy of atomization
of sodium = $+108\,\text{kJ mol}^{-1}$

ΔH_3 = standard enthalpy of atomization
of chlorine = $+121\,\text{kJ mol}^{-1}$

ΔH_4 = electron affinity of chlorine = $-364\,\text{kJ mol}^{-1}$

ΔH_5 = ionization energy of sodium = $+493\,\text{kJ mol}^{-1}$

ΔH_6 = lattice energy of sodium chloride

Substituting, $-411 = 108 + 121 - 364 + 494 + \Delta H_6$

$$\Delta H_6 = -770\,\text{kJ mol}^{-1}$$

The lattice energy of sodium chloride is $-770\,\text{kJ mol}^{-1}$.

N.B. Some scientists consider the lattice energy to be the endothermic change accompanying the reaction: $M^+X^-(s) \to M^+(g) + X^-(g)$.

Problems on Energetics

(Relative atomic masses will be found on page 132)

When required, assume that

$C(\text{graphite}) + O_2(g) \to CO_2(g) \quad \Delta H^\ominus = -394\,\text{kJ mol}^{-1}$
$H_2(g) + \tfrac{1}{2}O_2(g) \to H_2O(l) \quad \Delta H^\ominus = -286\,\text{kJ mol}^{-1}$

(1) Calculate the standard enthalpy of formation of ethanol given that its standard enthalpy of combustion is $-1400\,\text{kJ mol}^{-1}$.

(2) Calculate the standard enthalpy of formation of methane given that its standard enthalpy of combustion is $-895\,\text{kJ mol}^{-1}$.

(3) What is the standard enthalpy of combustion of methanol if its standard enthalpy of formation is $-239\,\text{kJ mol}^{-1}$?

(4) Calculate the standard enthalpy of formation of carbon monoxide, given that its standard enthalpy of combustion is $-284\,\text{kJ mol}^{-1}$.

(5) Calculate the standard enthalpy of formation of carbon disulphide, given: $\quad CS_2$

$S(s) + O_2(g) \to SO_2(g) \qquad \Delta H^\ominus = -294\,\text{kJ mol}^{-1}$
$CS_2(l) + 3O_2(g) \to CO_2(g) + 2SO_2(g) \quad \Delta H^\ominus = -1072\,\text{kJ mol}^{-1}$

(6) What is the standard enthalpy of combustion of ethoxyethane (diethyl ether) if its standard enthalpy of formation is $-280\,\text{kJ mol}^{-1}$?

(7) Calculate the standard enthalpy of formation of dimethylamine if its standard enthalpy of combustion is $-1760\,\text{kJ mol}^{-1}$.

(8) Calculate the standard enthalpy of formation of ethyne (acetylene) from the following data: C_2H_2
  (a) 2 g of carbon burned in excess oxygen liberated 66.1 kJ.
  (b) 2 g of hydrogen burned in excess oxygen liberated 284 kJ.
  (c) 2 g of ethyne burned in excess oxygen liberated 99.5 kJ.

(9) Calculate the standard enthalpy of hydrogenation of ethene (ethylene) from the data:

$C_2H_6(g) + 3\tfrac{1}{2}O_2(g) \rightarrow 2CO_2(g) + 3H_2O(l)$ $\Delta H^\ominus = -1550\,\text{kJ mol}^{-1}$
$C_2H_4(g) + 3O_2(g) \rightarrow 2CO_2(g) + 2H_2O(l)$ $\Delta H^\ominus = -1390\,\text{kJ mol}^{-1}$

(10) Calculate the standard enthalpy of formation of ethanal (acetaldehyde) given its standard enthalpy of combustion is $-1150\,\text{kJ mol}^{-1}$.

(11) Calculate the standard enthalpy of hydrogenation of ethyne (acetylene) to ethane from the following data:

  Standard enthalpy of combustion of ethane is $-1550\,\text{kJ mol}^{-1}$.
  Standard enthalpy of combustion of ethyne is $-1300\,\text{kJ mol}^{-1}$.

(12) Calculate the standard enthalpy of formation of benzene from the following data:

$$3C_2H_2(g) \rightarrow C_6H_6(l) \quad \Delta H^\ominus = -548\,\text{kJ mol}^{-1}$$

The standard enthalpy of combustion of ethyne (acetylene) is $-1300\,\text{kJ mol}^{-1}$.

(13) Calculate x in the reaction:

$$CuSO_4(s) + 5H_2O(l) \rightarrow CuSO_4.5H_2O \quad \Delta H^\ominus = x\,\text{kJ mol}^{-1}$$

By dissolving 1 g of anhydrous copper(II) sulphate in a large amount of water 418 kJ of heat are liberated. By dissolving 5 g of hydrated copper(II) sulphate in a large amount of water 230 kJ of heat are absorbed.

(14) 500 cm³ of 0.1 M sodium hydroxide solution at 15.50°C was quickly added to 500 cm³ of 0.1M nitric acid in a calorimeter which was also at a temperature of 15.50°C. The calorimeter had a mass of 540 g and a specific heat capacity of $0.390\,\text{J g}^{-1}\,°\text{C}^{-1}$. If the maximum temperature recorded was 16.15°C, calculate the enthalpy of neutralization. Assume the specific heat capacity of the solutions to be the same as that of water (i.e., $4.2\,\text{J g}^{-1}\,°\text{C}^{-1}$).

(15) Calculate the enthalpy of neutralization of potassium hydroxide by hydrochloric acid from the following data: 250 cm³ of 0.5M KOH at 12.0°C were mixed in a plastic beaker, of negligible heat capacity, with an equal volume of 0.5M HCl at the same temperature. The final temperature was 15.4°C. Assume the specific heat capacity of the solution to be $4.2\,J\,g^{-1}\,°C^{-1}$.

(16) 75 cm³ of 0.2M lithium hydroxide solution were added to 75 cm³ of 0.2M hydrochloric acid in a vacuum flask of water equivalent 10 g. The original temperature of the reactants before the reaction was 14.70°C and it rose to 16.00°C. Calculate the enthalpy of neutralization under these conditions.

(17) 100 cm³ of 0.5M ethanoic acid solution were mixed with 100 cm³ of 0.5M sodium hydroxide solution in a plastic beaker of negligible heat capacity. The original temperature of the reactants was 18.20°C and the final corrected temperature was 21.45°C. Calculate the enthalpy of neutralization of ethanoic acid by sodium hydroxide under these conditions. Assume the specific heat capacity of the solutions to be that of water, i.e., $4.2\,J\,g^{-1}\,°C^{-1}$.

(18) Calculate the standard enthalpy change for the thermit reaction between powdered aluminium and iron(III) oxide:

$$2Al(s) + Fe_2O_3(s) \rightarrow 2Fe(s) + Al_2O_3(s)$$

Given that,

$$2Al(s) + 1\tfrac{1}{2}O_2(g) \rightarrow Al_2O_3(s) \quad \Delta H^\ominus = -1600\,kJ\,mol^{-1}$$
$$2Fe(s) + 1\tfrac{1}{2}O_2(g) \rightarrow Fe_2O_3(s) \quad \Delta H^\ominus = -820\,kJ\,mol^{-1}$$

(19) Assuming the standard enthalpies of atomization of graphite and hydrogen to be respectively, $+715$ and $+218\,kJ\,mol^{-1}$ of gaseous atoms and that the standard enthalpy of formation of ethene (ethylene) is $+52\,kJ\,mol^{-1}$, calculate the standard enthalpy of formation of ethene from gaseous atoms of carbon and hydrogen. Then, using a value of $415\,kJ\,mol^{-1}$ for the mean C—H bond energy, calculate a value for the mean C═C bond energy.

(20) Using the mean bond energy values,

C—C $345\,kJ\,mol^{-1}$
C═C $610\,kJ\,mol^{-1}$
C—H $415\,kJ\,mol^{-1}$
H—H $435\,kJ\,mol^{-1}$

predict a value for the standard enthalpy change for the reaction:

$$C_2H_4(g) + H_2(g) \rightarrow C_2H_6(g)$$

(21) Using the mean bond energies,

C—C $345\,kJ\,mol^{-1}$
C—H $415\,kJ\,mol^{-1}$
C—O $340\,kJ\,mol^{-1}$
O—H $460\,kJ\,mol^{-1}$

calculate a value for the standard enthalpy of formation of ethanol from gaseous atoms of carbon, hydrogen and oxygen.

If the standard enthalpies of atomization of graphite, hydrogen and oxygen are respectively, $+715$, $+218$ and $+248\,\text{kJ}\,\text{mol}^{-1}$ of gaseous atoms, predict a value for the standard enthalpy of formation of ethanol from its elements in their normal states.

(22) Using the bond energies,

$$\begin{array}{ll} \text{C=C} & 610\,\text{kJ}\,\text{mol}^{-1} \\ \text{C—C} & 345\,\text{kJ}\,\text{mol}^{-1} \\ \text{C—H} & 415\,\text{kJ}\,\text{mol}^{-1} \\ \text{H—H} & 435\,\text{kJ}\,\text{mol}^{-1} \end{array}$$

and, assuming that the bonding between the carbon atoms of benzene consists of alternate double and single covalent bonds, predict a value for the standard enthalpy of hydrogenation of benzene and comment on the difference between this value and the experimentally determined value of approximately $-210\,\text{kJ}\,\text{mol}^{-1}$.

(23) Calculate the lattice energy of potassium chloride from the following data:

The standard enthalpy of formation of potassium chloride	$= -436\,\text{kJ}\,\text{mol}^{-1}$
The standard enthalpy of atomization of potassium	$= +90\,\text{kJ}\,\text{mol}^{-1}$
The standard enthalpy of atomization of chlorine	$= +121\,\text{kJ}\,\text{mol}^{-1}$
The electron affinity of chlorine	$= -364\,\text{kJ}\,\text{mol}^{-1}$
The ionization energy of potassium	$= +420\,\text{kJ}\,\text{mol}^{-1}$

(24) Calculate a value for the lattice energy of calcium chloride from the following data:

The standard enthalpy of formation of $CaCl_2(s)$	$= -795\,\text{kJ}\,\text{mol}^{-1}$
The standard enthalpy of atomization of calcium	$= +177\,\text{kJ}\,\text{mol}^{-1}$
The standard enthalpy of atomization of chlorine	$= +121\,\text{kJ}\,\text{mol}^{-1}$
The first ionization energy of calcium	$= +590\,\text{kJ}\,\text{mol}^{-1}$
The second ionization energy of calcium	$= +1100\,\text{kJ}\,\text{mol}^{-1}$
The electron affinity of chlorine	$= -364\,\text{kJ}\,\text{mol}^{-1}$

10

Distribution Constant

Theory

The *Distribution Law* states that: If a solute X distributes itself between two immiscible solvents, A and B, at the same temperature, and X is in the same molecular condition in both A and B,

$$\frac{\text{Concentration of } X \text{ in } A}{\text{Concentration of } X \text{ in } B} = \text{a constant}$$

This constant is called *Distribution Constant* of X between A and B. In the limit, when saturation is reached, the distribution constant is equal to:

$$\frac{\text{Solubility of } X \text{ in } A}{\text{Solubility of } X \text{ in } B}$$

where solubility is expressed in grams per unit volume. (A more rigorous approach considers the concentrations initially to be in mol dm^{-3}.)

Examples

(1) *The following figures relate to iodine in tetrachloromethane (carbon tetrachloride) and water when the solutions are in equilibrium at a certain temperature:*

	grams of iodine per dm^3			
In CCl_4	17.5	15.8	13.2	6.50
In H_2O	0.200	0.180	0.150	0.075

Calculate the distribution constant of iodine between these two solvents. What is the effect of shaking 1 dm^3 of water containing 0.3 g of iodine with 10 cm^3 of tetrachloromethane?

The ratios of the concentrations given are:

$$\frac{17.5}{0.200} \qquad \frac{15.8}{0.180} \qquad \frac{13.2}{0.150} \qquad \frac{6.50}{0.075}$$
$$= 87.5 \qquad = 87.8 \qquad = 88.0 \qquad = 86.7$$

These figures are approximately constant and the average of them is 87.5. This is the required distribution constant.

Let x g of iodine pass into the 10 cm^3 of tetrachloromethane. Then, $(0.3 - x)$ g remain in the water. By the distribution law,

$$\frac{\frac{x}{10}}{\frac{(0.3 - x)}{1000}} = \frac{87.5}{1}$$

Therefore, $\qquad \dfrac{x}{10} = \dfrac{87.5(0.3 - x)}{1000}$

From this, $\qquad x = (0.875 \times 0.3) - 0.875x$

$\qquad\qquad\qquad = 0.2625 - 0.875x$

so that $\qquad x = \dfrac{0.2625}{1.875} = 0.14$

That is, 0.14 g of iodine enter the tetrachloromethane, while 0.16 g of iodine remain in water.

(2) *100 cm^3 of water contain 10 g of an organic compound, X. If the mixture is shaken with (a) 50 cm^3 of ethoxyethane, (b) 25 cm^3 of ethoxyethane twice, what mass of X is extracted? The distribution constant of X between ethoxyethane and water is 9. Assume that ethoxyethane and water are immiscible and that X is in the same molecular condition in both.*

(a) Let a g of X pass into the ethoxyethane, leaving $(10 - a)$ g in the water. Then, by the distribution law,

$$\frac{\frac{a}{50}}{\frac{(10 - a)}{100}} = \frac{9}{1}$$

Therefore, $\qquad \dfrac{a}{50} = \dfrac{9(10 - a)}{100}$

or $\qquad 2a = 90 - 9a$

and $\qquad a = \dfrac{90}{11} = 8.18 \text{ g}$

That is, 8.18 g of X are extracted by the ethoxyethane.

(b) *First Extraction*

Let b g of X pass into ethoxyethane, leaving $(10 - b)$ g in water. Then, by the distribution law,

$$\frac{\frac{b}{25}}{\frac{(10-b)}{100}} = \frac{9}{1}$$

Solving this as above, $\quad b = 6.92\,\text{g}$

Second Extraction

The first extraction leaves $(10 - 6.92)$ g, or 3.08 g, of X in water. The conditions are the same again, so that $\frac{6.92}{10} \times 3.08$ g of X must pass into ethoxyethane in the second extraction. This is 2.13 g. The total extraction is $(6.92 + 2.13)$ g or 9.05 g, of X into ethoxyethane. Notice that it is more effective to use the ethoxyethane in two small extractions than in a single large one.

The simple Distribution Law assumes the solute X to be in the same molecular condition in both solvents. If it is polymerized as X_n in solvent A and exists as simple X in solvent B, the law takes the form:

$$\frac{\sqrt[n]{\text{Concentration of } X \text{ in } A}}{\text{Concentration of } X \text{ in } B} = \text{a constant}$$

(A and B are again assumed immiscible and the temperature constant.)

The best known cases of this form of the law occur with benzoic acid and ethanoic acid, which occur as dimers in benzene solution. The following figures illustrate the case of benzoic acid.

Grams of benzoic acid per dm^3

In benzene	24.2	31.2	41.2	50.9
In water	1.50	1.70	1.95	2.13

If benzoic acid is dimerized in benzene, the ratios:

$$\frac{\sqrt{24.2}}{1.50} \quad \frac{\sqrt{31.2}}{1.70} \quad \frac{\sqrt{41.2}}{1.95} \quad \frac{\sqrt{50.9}}{2.13} \quad \text{should be constant}$$

They yield the figures: 3.28, 3.29, 3.29 and 3.35 respectively. These are approximately constant, giving evidence of the existence of benzoic acid as the dimer, $(C_6H_5COOH)_2$, in benzene.

Problems on Distribution Constant

(*Relative atomic masses will be found on page* 132)

(1) A mass of butanedioic (succinic) acid known to be equivalent to $62.0\,\text{cm}^3$ of a certain sodium hydroxide solution was thoroughly shaken with equal volumes of ethoxyethane and water. The water layer required $9.8\,\text{cm}^3$ of the sodium hydroxide solution.

The experiment was repeated several times, increasing the mass of the acid by one-third of the original mass each time. The volumes of sodium hydroxide required for the water layers were:

(2)	. . .	$12.8\,\text{cm}^3$
(3)	. . .	$16.6\,\text{cm}^3$
(4)	. . .	$20.0\,\text{cm}^3$
(5)	. . .	$23.0\,\text{cm}^3$

Calculate the mean value of the distribution constant of butanedioic acid between ethoxyethane and water (to three significant figures).

(2) Iodine was shaken with equal volumes of tetrachloromethane and water. Calculate the mean distribution constant of iodine between these two solvents from the results:

	I	II	III
$25\,\text{cm}^3$ CCl_4 layer required cm^3 0.1M thiosulphate	27.7	21.2	14.0
$25\,\text{cm}^3$ H_2O layer required cm^3 0.01M thiosulphate	3.15	2.4	1.6

What mass of iodine would dissolve in the water if $5.00\,\text{g}$ of iodine were taken with $1\,\text{dm}^3$ of tetrachloromethane and $2\,\text{dm}^3$ of water?

(3) A solid S is soluble in two immiscible liquids A and B, and it is in the same molecular condition in both. Also,

$$\frac{\text{Solubility of S in A}}{\text{Solubility of S in B}} = \frac{10}{1}$$

$5\,\text{g}$ of S were dissolved in $100\,\text{cm}^3$ of B and extracted with $100\,\text{cm}^3$ of A. What mass of S remained in the B layer? What mass of solid would have remained in the B layer if the extraction had been carried out with two separate portions of $50\,\text{cm}^3$ of A?

(4) $1.00\,\text{g}$ of iodine was shaken with $50\,\text{cm}^3$ of tetrachloromethane and $50\,\text{cm}^3$ of water. $25\,\text{cm}^3$ of the aqueous layer required $4.5\,\text{cm}^3$ 0.01M sodium thiosulphate solution. Calculate the distribution constant of iodine between these two solvents.

(5) $10\,\text{g}$ of a solid, X, are dissolved in $1\,\text{dm}^3$ of a solvent, Y, and the solute is extracted by a liquid Z such that

$$\frac{\text{Solubility of X in Z}}{\text{Solubility of X in Y}} = \frac{8}{1}$$

Calculate, to the nearest integer, the percentage of X left in the solvent Y if it is extracted with:

(a) 1 dm³ of Z.
(b) ½ dm³ of Z twice.
(c) ⅓ dm³ of Z three times.

(6) Show, from the following results, that ethanoic acid is associated into double molecules in benzene solution.

Distribution of acid between benzene and water.

Molar concentration in benzene	Molar concentration in water
0.027	0.22
0.072	0.36
0.128	0.48
0.25	0.67

(7) Show that the following results are in accordance with the belief that benzoic acid is associated in benzene solution into double molecules.

Distribution of benzoic acid between benzene and water.

Molar concentration in benzene	Molar concentration in water
0.252	0.0150
0.375	0.0183
0.509	0.0213
0.817	0.0270

(8) A crude sample of lead contained 2% of silver by mass. What mass of silver would be left in 1000 kg of the lead if it were thoroughly agitated with 50 kg of zinc at 800°C and the zinc removed? (At 800°C the solubility of silver in a given mass of zinc is 300 times its solubility in an equal mass of lead.)

11

Electrolysis

Theory

Most calculations on electrolysis are applications of Faraday's Laws of Electrolysis, the first of which states that:

> The mass of substance liberated by electrolysis is directly proportional to the quantity of electricity which is passed.

The quantity of electricity is measured in coulombs, where:

> Number of coulombs = current in amperes × time in seconds.

1 mole of singly charged ions will require to either gain or lose 1 mole of electrons for neutralization; similarly, 1 mole of doubly charged ions will require 2 moles of electrons. The transfer of 1 mole of electrons corresponds to the passage of approximately 96 500 coulombs, this giving the Faraday constant. This leads to the following method of stating Faraday's second law:

> When the same quantity of electricity is passed through different electrolytes, the masses of the different substances liberated are directly proportional to the relative atomic masses of the substances divided by their ionic charges.

These two laws can be summed up in terms of the Faraday constant. The Faraday constant is the quantity of electricity that passes when one mole of a univalent element (or half a mole of a divalent element, etc.) is discharged or dissolved in electrolysis.

The electrochemical equivalent of an element is the mass of the element, in grams, which is liberated by the passage of 1 coulomb of electricity, i.e., 1 ampere for 1 second.

Examples

(1) *0.198 g of copper are deposited on a cathode in 40 minutes by passing a steady current of 0.25 ampere through copper(II) sulphate solution. Calculate the electrochemical equivalent of copper.*

The electrochemical equivalent of copper is the mass of it deposited by a current of one ampere passing for 1 second. On the above figures, it is:

$$0.198 \times \frac{1}{40 \times 60} \times \frac{1}{0.25} = 0.000330 \text{ g C}^{-1}$$

(2) *The electrochemical equivalent of silver is 0.00112 g C^{-1}. What mass of silver is deposited by the passage of a steady current of 0.5 ampere for 1 hour in a silver-plating bath?*

Number of coulombs $= 0.5 \times 60 \times 60$

Mass of silver $\quad\quad = 0.5 \times 60 \times 60 \times 0.00112 = 2.016 \text{ g}$

(3) *A steady current of 0.65 ampere is passed for 5.5 hours through solutions of sulphuric acid and copper(II) sulphate in series. Calculate the volumes of hydrogen and oxygen liberated at 13°C and $100\,000 \text{ N m}^{-2}$ (750 mmHg) pressure from the acid, and the mass of copper precipitated.*

As the hydrogen ion carries a single positive charge (i.e., as H^+ or $H^+(aq)$), 1 mole of hydrogen ions will require 1 mole of electrons, i.e., 96 500 coulombs, for conversion into neutral atoms.

1 mole of hydrogen atoms will produce 0.5 mole of hydrogen gas (H_2), which will occupy 11.2 dm^3 at s.t.p., i.e., 96 500 coulombs will liberate 11.2 dm^3 of hydrogen at s.t.p.

The passage of 0.65 ampere for 5.5 hours represents $(0.65 \times 5.5 \times 60 \times 60)$ coulombs, or 12 870 coulombs.

12 870 coulombs will liberate $11.2 \times \dfrac{12\,870}{96\,500}$ dm^3 of H_2 at s.t.p.

$= 11.2 \times \dfrac{12\,870}{96\,500} \times \dfrac{286}{273} \times \dfrac{101\,300}{100\,000}$ at 13°C and $100\,000 \text{ N m}^{-2}$

$= 1.585 \text{ dm}^3$

Oxygen is liberated from OH^- ions according to the electrode reactions:

$$OH^- \rightarrow OH + e^-$$
$$4OH \rightarrow 2H_2O + O_2$$

Hence 1 mole of electrons will be released by 1 mole of OH^- ions, which will produce 0.25 mole of oxygen (O_2). This is half the volume of hydrogen liberated by 96 500 coulombs, therefore, the volume of oxygen liberated by 12 870 coulombs = $\frac{1.585}{2}$ = 0.7925 dm^3 at 13°C and 100 000 N m^{-2}.

The copper(II) ion carries a double positive charge, therefore 1 mole of electrons (96 500 coulombs) will liberate 0.5 mole of copper = $\frac{64}{2}$ g.

12 870 coulombs will liberate $\frac{64}{2} \times \frac{12\,870}{96\,500}$ = 4.268 g

Problems on Electrolysis

(Relative atomic masses will be found on page 132)

(1) 0.792 g of copper are deposited in eighty minutes by a steady current of half an ampere passed through a solution of copper(II) sulphate. Calculate the electrochemical equivalent of copper(II).

(2) The electrochemical equivalent of silver is 0.00112 g C^{-1}. What mass of silver will be deposited in a silver-plating bath by passing a current of 1.5 ampere for an hour and a half?

(3) A steady current of 0.27 ampere passed for half an hour in a water voltameter liberated 56 cm^3 of hydrogen at s.t.p. Calculate the electrochemical equivalent of hydrogen. What are the electrochemical equivalents of (*a*) oxygen, (*b*) aluminium, (*c*) zinc.

(4) If the cost of electricity to produce magnesium is £x kg^{-1} of the metal, what is the cost of electricity for producing y kg of aluminium at the same rates?

(5) What volume of hydrogen at 15°C and 100 700 N m^{-2} (755 mmHg) would be liberated in a water voltameter by the passing of half an ampere for 30 minutes?

(6) An electric current is passed through solutions of copper(II) sulphate, silver nitrate and sulphuric acid, in series. If 0.21 g of copper are deposited in the first cell, calculate (*a*) the mass of silver deposited in the second, (*b*) the volume of hydrogen liberated in the third at 15°C and 98 640 N m^{-2} (740 mmHg).

(7) An electric current is passed in series through a water voltameter and a solution of silver nitrate. 56 cm^3 of hydrogen are liberated in the voltameter, measured wet at 16°C and 100 000 N m^{-2} (750 mmHg). Calculate the mass of silver deposited at the cathode in the other cell. (The vapour pressure of water at 16°C is 4330 N m^{-2} (14 mmHg).)

12

Equilibria

Theory

In this section, and in following sections, it is sometimes necessary to solve quadratic equations in the course of chemical calculations. This is done by application of the algebraic formula which states that, if $ax^2 + bx + c = 0$,

$$x = \frac{-b \pm \sqrt{b^2 - 4ac}}{2a}$$

It is expected that students can apply this formula as required and full solutions are not given in the following examples.

In general, the Equilibrium Law is applied to homogeneous systems in equilibrium, i.e., to systems involving reagents which are either entirely liquid or entirely gaseous. If special assumptions are made, the Law can be applied to other cases, some of which are mentioned later.

Consider the homogeneous system:

$$wA + xB \rightarrow yC + zD$$

then, when the system is in a state of equilibrium, the equilibrium molar concentrations of reactants and products are related in the following way,

$$\frac{[C]^y[D]^z}{[A]^w[B]^x} = K_c$$

which, for a fixed temperature, is a constant. This relationship is the Equilibrium Law and K_c is known as the equilibrium constant.

Examples

(1) *It is found that, if 1 mole of ethanoic acid (acetic acid) and $\frac{1}{2}$ mole of ethanol react to equilibrium at a certain temperature, 0.422 moles of ethyl ethanoate (ethyl acetate) are produced. Show that the equilibrium constant of the reaction is about 4. Using this integral value for the constant, calculate the composition of the equilibrium mixture obtained if the following are allowed to react to equilibrium at the same temperature, (a) 3 moles of ethanoic acid and 5 moles of ethanol, (b) 3 moles of ethanoic acid, 5 moles of ethanol and 2.5 moles of water.*

The reaction is:

$$CH_3COOH + C_2H_5OH \rightleftharpoons CH_3COOC_2H_5 + H_2O$$

For the first case, originally

    1.000   0.500   —   — mole

at equilibrium

(1.000 − 0.422) (0.500 − 0.422) 0.422   0.422 mole
= 0.578   = 0.078

If the volume of the mixture is V dm^3, the equilibrium concentrations are,

$$\frac{0.578}{V} \quad \frac{0.078}{V} \quad \frac{0.422}{V} \quad \frac{0.422}{V} \quad \text{mol dm}^{-3}$$

The equilibrium constant

$$K_c = \frac{[CH_3COOC_2H_5][H_2O]}{[CH_3COOH][C_2H_5OH]}$$

$$= \frac{0.422}{V} \times \frac{0.422}{V} \times \frac{V}{0.578} \times \frac{V}{0.078}$$

$$= 3.95$$

i.e., the equilibrium constant is about 4.

(a) Let x mole of ester and water be formed at equilibrium. The equilibrium amounts are therefore,

$$CH_3COOH + C_2H_5OH \rightleftharpoons CH_3COOC_2H_5 + H_2O$$
$$(3-x) \qquad (5-x) \qquad\qquad x \qquad\qquad x \text{ mole}$$

Applying the Equilibrium Law as in the case above,

$$K_c = 4 = \frac{x^2}{(3-x)(5-x)}$$

This expression simplifies to $3x^2 - 32x + 60 = 0$. Solved by the formula (p. 85), it gives $x = 2.42$. The equilibrium mixture contains acid 0.58, alcohol 2.58, ester 2.42 and water 2.42 mole.

(b) Let x mole of water and ester be formed at equilibrium. The position is then:

$$CH_3COOH + C_2H_5OH \rightleftharpoons CH_3COOC_2H_5 + H_2O$$
$$(3-x) \quad\quad (5-x) \quad\quad\quad x \quad\quad\quad (2.5+x) \text{ mole}$$

Applying the Equilibrium Law,

$$\frac{x(2.5+x)}{(3-x)(5-x)} = K_c = 4$$

This simplifies to: $3x^2 - 34.5x + 60 = 0$. Solved by the formula (p. 87), it gives $x = 2.13$. The equilibrium mixture is acid 0.87, alcohol 2.87, ester 2.13, water 4.63 mole.

(2) *At a temperature close to 400°C, hydrogen iodide has a degree of dissociation of 20%. Calculate the composition of the equilibrium mixture produced if 1 mole of hydrogen and 2.1 mole of iodine react to equilibrium at this temperature*

The dissociation of hydrogen iodide is represented by the equation:

$$2HI \rightleftharpoons H_2 + I_2$$
At equilibrium $\quad (1-0.2) \quad 0.1 \quad 0.1 \quad$ mole

These figures state the fact that, for every 1 mole of HI originally present, 20% or 0.2 mole has dissociated. The dissociation of a certain number of HI molecules produces half that number of iodine and hydrogen molecules. If the volume of the mixture at equilibrium is V dm³, the equilibrium concentrations are,

$$\frac{0.8}{V} \quad \frac{0.1}{V} \quad \frac{0.1}{V} \quad \text{mol dm}^{-3}$$

$$K_c = \frac{[H_2][I_2]}{[HI]^2} = \frac{\frac{0.1}{V} \times \frac{0.1}{V}}{\left(\frac{0.8}{V}\right)^2}$$

Notice that the concentration of HI is squared because two molecules of it are operative.

From this

$$K_c = \frac{(0.1)^2}{(0.8)^2} = \frac{1}{64} = 1.56 \times 10^{-2}$$

Notice that the constant is independent of V and, therefore, the pressure at which the reaction is carried out.

In the second case, let $2x$ mole of hydrogen iodide be formed. Then we have at equilibrium:

$$2HI \rightleftharpoons H_2 + I_2$$
$$2x \quad (1-x) \quad (2.1-x) \text{ mole}$$

Applying the Equilibrium Law as above,

$$\frac{(1-x)(2.1-x)}{(2x)^2} = \frac{1}{64}$$

This simplifies to: $15x^2 - 49.6x + 33.6 = 0$. Solved by the formula (p. 87), it gives $x = 0.95$. The equilibrium mixture contains hydrogen 0.05, iodine 1.15, hydrogen iodide 1.90 mole.

(3) *At 1 atmosphere pressure, and 55°C, N_2O_4 is 50% dissociated. Calculate the equilibrium constant for this reaction in terms of pressure. Hence calculate the degree of dissociation of the gas at 55°C and 10 atmospheres pressure.*

$$N_2O_4 \rightleftharpoons 2NO_2$$
At equilibrium $(1 - 0.5)$ 1.0 mole

Total number of moles of gas present at equilibrium

$$= (1 - 0.5) + 1.0 = 1.5$$

The fraction of these which are $N_2O_4 = \dfrac{0.5}{1.5} = 0.33$

The partial pressure of N_2O_4 at equilibrium

$$= \text{total pressure} \times 0.33$$
$$= 1 \times 0.33 = 0.33 \,\text{atm} = p_{N_2O_4}$$

Similarly the partial pressure of $NO_2 = 1 \times \dfrac{1.0}{1.5} = 0.67\,\text{atm} = p_{NO_2}$

By applying the Equilibrium Law and the fact that partial pressures are directly proportional to molar concentrations, an equilibrium constant

$$K_p = \frac{(p_{NO_2})^2}{p_{N_2O_4}} = \frac{(0.67)^2}{0.33} = \frac{4}{3}\text{atm}$$

If, at 10 atmospheres pressure, the degree of dissociation is α, the situation is:

$$N_2O_4 \rightleftharpoons 2NO_2$$
$(1 - \alpha)$ 2α mole

The total number of moles present is $(1 + \alpha)$, so the partial pressures are:

$$N_2O_4 \quad \frac{1-\alpha}{1+\alpha} \times 10\,\text{atm}$$

$$NO_2 \quad \frac{2\alpha}{(1+\alpha)} \times 10\,\text{atm}$$

Applying the Equilibrium Law,

$$\frac{\left(\frac{2 \times 10}{1+\alpha}\right)^2}{\frac{10(1-\alpha)}{1+\alpha}} = K_p = \frac{4}{3}$$

this simplifies to $124\alpha^2 = 4$, from which it is found that $\alpha = 0.18$, i.e., in the conditions stated, N_2O_4 is 18% dissociated.

(4) *When the Equilibrium Law is applied to heterogeneous systems, the concentrations of pure solids are regarded as being constant and of unit value.*

Example

At a certain temperature, the equilibrium pressures of steam and hydrogen in contact with iron and its black oxide, Fe_3O_4, were found to be $1533 \, N \, m^{-2}$ ($11.5 \, mmHg$) and $32\,190 \, N \, m^{-2}$ ($241.5 \, mmHg$) respectively. Calculate (a) the pressure of hydrogen in equilibrium with $1066 \, N \, m^{-2}$ ($8 \, mmHg$) of steam pressure, (b) the pressures of hydrogen and steam at a total pressure of $101\,300 \, N \, m^{-2}$ ($760 \, mmHg$) at this temperature.

The reaction involved is:

$$3Fe + 4H_2O \rightleftharpoons Fe_3O_4 + 4H_2$$

Provided both are present in some amount, the concentrations of Fe and Fe_3O_4 (solids) can be taken as constant.

By applying the Equilibrium Law,

$$K_c = \frac{[H_2]^4}{[H_2O]^4}$$

or if partial pressures are used,

$$\frac{(p_1)^4}{(p_2)^4} = \text{a constant},$$

where p_1 and p_2 are the partial pressures of hydrogen and steam respectively.

Taking the fourth root of this expression,

$$\frac{p_1}{p_2} = \text{a constant} = K_p$$

From the given figures,

$$K_p = \frac{32\,190}{1533} = 21.0$$

(*a*) If the equilibrium pressure of steam is to be $1066 \, N \, m^{-2}$,

$$\frac{p_1}{1066} = K_p = 21.0$$

$$p_1 = 22\,390 \, N \, m^{-2}$$

(b) Let the steam pressure be x N m^{-2} at equilibrium. Then the hydrogen pressure is $(101\,300 - x)$ N m^{-2}. This gives:

$$\frac{101\,300 - x}{x} = 21.0$$

Therefore $\qquad\qquad\qquad x = 4600$ N m^{-2}

From this, the equilibrium pressures are:

$\qquad\qquad$ hydrogen $(101\,300 - 4600) = 96\,700$ N m^{-2}
$\qquad\qquad$ steam $= 4600$ N m^{-2}

Problems on the Equilibrium Law
(Relative atomic masses will be found on page 132)

(1) If a mixture of ethanol and pure ethanoic acid (glacial acetic acid) in equi-molar proportions is allowed to react until equilibrium is attained at 25°C, two-thirds of the acid is esterified. Calculate the equilibrium constant of the reaction:

$$C_2H_5OH + CH_3COOH \rightleftharpoons CH_3COOC_2H_5 + H_2O$$

Hence calculate the molar composition of the mixtures reached at equilibrium at 25°C, starting from (a) 1 mole of ethanoic acid and 0.5 mole of ethanol, (b) 1 mole of ethanoic acid and 8 mole of ethanol, (c) 1 mole each of ethanoic acid, ethanol and water.

(2) In an experiment of Bodenstein, 20.57 mole of hydrogen and 5.22 mole of iodine were allowed to react to equilibrium at 450°C. At this point, the mixture contained 10.22 mole of hydrogen iodide. Show, from the given data, that the equilibrium constant of the reaction

$$H_2 + I_2 \rightleftharpoons 2HI$$

is 61 to the nearest integer.

If, in a similar experiment, 20 mole of hydrogen and 60 mole of iodine are used, calculate the molar composition of the equilibrium mixture at 450°C. (Use $K = 61$.)

(3) 2.00 g of phosphorus pentachloride are allowed to reach equilibrium at 200°C in a vessel of 1 dm^3 capacity. If the equilibrium constant of the reaction

$$PCl_5 \rightleftharpoons PCl_3 + Cl_2$$

is 0.008 mol dm^{-3} at this temperature and in the conditions stated, calculate the percentage dissociation of the phosphorus pentachloride at equilibrium.

(4) Solid ammonium hydrogensulphide dissociates on slight warming according to the equation
$$NH_4HS \rightleftharpoons NH_3 + H_2S$$
At a certain temperature, the system is in equilibrium, with the partial pressures of the two gases equal at $33\,330\text{ N m}^{-2}$ (250 mmHg). What is the partial pressure of ammonia if hydrogen sulphide is added till its partial pressure is $40\,000\text{ N m}^{-2}$ (300 mmHg) without change of volume or temperature? Under what conditions would the partial pressure of hydrogen sulphide be $20\,000\text{ N m}^{-2}$ (150 mmHg) at the same temperature and volume?

(5) At a certain temperature, and at a pressure of 1 atmosphere, iodine vapour contains 40% by volume of iodine atoms.
$$I_2 \rightleftharpoons I + I$$
At what total pressure (without temperature change) would this percentage be reduced to 20?

(6) At a certain high temperature, the equilibrium constant of the reaction $N_2 + O_2 \rightleftharpoons 2NO$ is 8×10^{-4}. Assuming air to be a mixture of four volumes of nitrogen with one volume of oxygen, calculate the percentage of nitrogen monoxide by volume in the gas produced by allowing air to reach equilibrium at this temperature.

(7) The equilibrium constant of the reaction
$$CO_2 + H_2 \rightleftharpoons CO + H_2O$$
at 1000°C is 1.6. What is the percentage by volume of each gas in the equilibrium mixture at 1000°C produced from (*a*) a mixture of 50 cm³ hydrogen and 50 cm³ carbon dioxide, (*b*) 25 cm³ each of carbon dioxide and monoxide and 50 cm³ of hydrogen.

(8) Haber's synthesis of ammonia depends on the reaction
$$N_2 + 3H_2 \rightleftharpoons 2NH_3$$
Show that, if the partial pressure of ammonia remains small, a doubling of the total pressure raises K_p fourfold.

IONIC EQUILIBRIA

Theory

WEAK ELECTROLYTES (DILUTION LAW)

In its simplest form, the Dilution Law states the behaviour of a weak electrolyte with respect to ionization in dilute solutions of varying concentrations. It is an application of the Equilibrium Law in the ionic field.

If the degree of ionization of a binary electrolyte, HA, is α at a dilution of 1 mole of the electrolyte in V dm^3 of solution at a constant temperature, the Dilution Law is determinated as follows:

The equilibrium is represented by,

$$HA \rightleftharpoons H^+ + A^-$$

Equilibrium
amounts, $\qquad\qquad (1-\alpha) \qquad \alpha \qquad \alpha$ mole

Equilibrium concentrations,

$$\frac{(1-\alpha)}{V} \qquad \frac{\alpha}{V} \qquad \frac{\alpha}{V} \text{ mol dm}^{-3}$$

By applying the Equilibrium Law,

$$K = \frac{[H^+][A^-]}{[HA]} = \frac{\alpha^2}{V(1-\alpha)}$$

This statement is known as the Dilution Law for a binary electrolyte and K is called the Dissociation Constant of the electrolyte. If an electrolyte, which dissociates into more than two ions, is being considered it is necessary to redetermine the formula.

If the electrolyte is very weak, so that α is small, and $(1-\alpha)$ can be taken as equal to 1, the above formula simplifies to:

$$K = \frac{\alpha^2}{V}, \text{ so that } \alpha^2 = KV \text{ and } \alpha = \sqrt{KV}$$

As K is a constant for a constant temperature, this means that, at varying dilutions, α, the degree of ionization of a very weak electrolyte, is directly proportional to the square root of the volume of solution containing 1 mole of the electrolyte.

K is sometimes expressed on a logarithmic scale, which, for an acid, is denoted by pK_a, where p$K_a = -\log K_a$ and for a base by pK_b, where p$K_b = -\log K_b$.

HYDROGEN ION INDEX, pH

The hydrogen ion index, pH, is a measure of the acidity or alkalinity of a solution and is defined in the following way:

If the hydrogen ion concentration in a solution is 10^{-x} mol dm^{-3}, the pH of the solution is x, i.e.,

$$\text{pH} = -\log[H^+]$$

$[H^+]$ is measured in mol dm^{-3} and log represents logarithm to base 10. In water at 25°C, the following situation exists,

$$K_w = [H^+][OH^-] = 1 \times 10^{-14} \, mol^2 \, dm^{-6}$$

where K_w is a constant called the ionic product of water. It follows from this that,

$$pH + pOH = 14 \quad \text{or} \quad pH = 14 - pOH$$

This fixed relation makes it possible to express alkalinity (caused by OH^-), as well as acidity, in terms of pH. To summarize the situation at 25°C,

pH = 7 represents neutrality (the state of pure water in which $[H^+] = [OH^-] = 10^{-7} \, mol \, dm^{-3}$)

pH less than 7 represents acidity ($[H^+]$ greater than $[OH^-]$)

pH greater than 7 represents alkalinity ($[H^+]$ less than $[OH^-]$)

Notice that, because pH is really a logarithmic index from which the negative sign has been dropped by convention, a *low* pH represents a *high* level of acidity and a *falling* pH represents a *rising concentration of* H^+.

Examples

(1) *The degree of ionization of ethanoic acid (acetic acid) in 0.1 M solution at a certain temperature is 0.014. Calculate the dissociation constant of the acid and its* pK_a *value at this temperature. What is the degree of ionization and the pH of a 0.01M solution of ethanoic acid at this temperature?*

Using the Dilution Law, as previously derived for a monobasic acid,

$$K_a = \frac{\alpha^2}{V(1-\alpha)}$$

where $\alpha = 0.014$, $(1-\alpha) = 0.986$ and $V = 10 \, dm^3$. That is

$$K_a = \frac{(0.014)^2}{10 \times 0.986} = 1.99 \times 10^{-5}$$

$$\log K_a = \overline{5}.2989 = -5 + 0.2928 = -4.7011$$

$$pK_a = -\log K_a = 4.70$$

In the case of the 0.01 M acid $V = 100 \, dm^3$. Then if the degree of ionization is α_1

$$K_a = 1.99 \times 10^{-5} = \frac{\alpha_1^2}{100(1-\alpha_1)}$$

This yields the equation: $\alpha_1^2 + 0.00199\alpha_1 - 0.00199 = 0$, and this, when solved by the formula (p. 87), gives $\alpha_1 = 0.0436$.

Using the approximate form of the Dilution Law, where $(1 - \alpha_1)$ is taken as being equal to 1,

$$K_a = 1.99 \times 10^{-5} = \frac{\alpha_1^2}{100} \quad \text{and} \quad \alpha_1 = 0.0442$$

If the acid is fully ionized in 0.01M solution,

$$[H^+] = 1 \times 10^{-2}\,\text{mol dm}^{-3}$$

Since the degree of ionization (to two significant figures) is 0.044, the actual situation is:

$$[H^+] = (1 \times 10^{-2} \times 0.044) = 4.4 \times 10^{-4}\,\text{mol dm}^{-3}$$
$$\log[H^+] = \bar{4}.64 = -4 + 0.64 = -3.36$$
$$pH = -\log[H^+] = 3.36$$

That is, the pH of the 0.1M solution is about 3.4.

(2) *The dissociation constant of ammonia is $1.8 \times 10^{-5}\,mol\,dm^{-3}$. Calculate the pH of decimolar ammonia solution. Also calculate the pH of the solution obtained by adding 2.5 g of ammonium chloride (assumed fully ionized) to 250 cm^3 of decimolar ammonia solution.*

Let the degree of ionization of the 0.1M ammonia solution be α. The dilution, V (the volume in dm^3 containing 1 mole of ammonia), is $10\,dm^3$. The ionization can be represented by the equation,

$$H_2O + NH_3 \rightleftharpoons NH_4^+ + OH^-$$

At equilibrium $\qquad\qquad (1 - \alpha) \qquad \alpha \qquad \alpha$ moles

Equilibrium concentrations

$$\frac{(1 - \alpha)}{10} \qquad \frac{\alpha}{10} \qquad \frac{\alpha}{10} \quad \text{mol dm}^{-3}$$

Therefore,

$$K_b = 1.8 \times 10^{-5} = \frac{\alpha^2}{10(1 - \alpha)} \,\text{mol dm}^{-3}$$

This yields the equation: $\alpha^2 + 0.00018\alpha - 0.00018 = 0$, which, when solved by the formula (p. 85), gives $\alpha = 0.0135$.

The concentration of OH^- in the solution would be $1 \times 10^{-1}\,\text{mol dm}^{-3}$ if the electrolyte was fully ionized. In the actual case,

$$[OH^-] = (1 \times 10^{-1}) \times 0.0135 = 1.35 \times 10^{-3}\,\text{mol dm}^{-3}$$
$$\log[OH^-] = \bar{3}.13 = -3 + 0.13 = -2.87$$
$$pOH = -\log[OH^-] = 2.87$$

That is, pOH is about 2.9 and pH is $(14 - 2.9)$ or 11.1.

The ammonium chloride added is $10\,\text{g}\,\text{dm}^{-3}$, which is $\frac{10}{53.5}$, or 0.187 $\text{mol}\,\text{dm}^{-3}$. If it is fully ionized, the ammonium ion concentration from this source is $1.87 \times 10^{-1}\,\text{mol}\,\text{dm}^{-3}$. The contribution of ammonia to this concentration is relatively negligible when equilibrium is established. From the Equilibrium Law,

$$\frac{[NH_4^+][OH^-]}{[NH_3]} = K_b = 1.8 \times 10^{-5}\,\text{mol}\,\text{dm}^{-3}$$

That is, $\quad \dfrac{(1.87 \times 10^{-1})[OH^-]}{(1 \times 10^{-1})} = 1.8 \times 10^{-5}\,\text{mol}\,\text{dm}^{-3}$

This yields,

$$[OH^-] = 9.6 \times 10^{-6}\,\text{mol}\,\text{dm}^{-3}$$
$$\log[OH^-] = \bar{6}.98 = -6 + 0.98 = -5.02$$
$$-\log[OH^-] = 5.02$$

That is pOH is about 5 and pH is about $(14 - 5)$ or 9.

Problems on the Dilution Law and pH

(*Relative atomic masses will be found on page* 132)

(9) Show that the following figures for ethanoic acid (acetic acid) at 25°C approximately satisfy the Dilution Law:

Vol. in dm^3 containing 1 mol of the acid	Degree of ionization of the acid
5·4	0.0098
10.6	0.0138
24.9	0.0212
63.3	0.0336

(10) The dissociation constant of methylamine at 25°C is 5×10^{-4} $\text{mol}\,\text{dm}^{-3}$. Calculate the hydroxide concentration in a 0.1M solution of this base and hence the pH of the solution.

(11) Aqueous ammonia is 1.4% ionized at 25°C in 0.1M solution. Calculate the dissociation constant of the base and its pK_b value at this temperature. Obtain an approximate value for its degree of ionization in 0.01M solution. What is the hydroxide ion concentration of this solution and hence its pH?

(12) The ionic product for water is $1 \times 10^{-14}\,\text{mol}^2\,\text{dm}^{-6}$ at 25°C. Calculate the pH of the following solutions, assuming complete ionization of the solutes unless otherwise stated: (*a*) 0.01M hydrochloric acid, (*b*) 0.001M potassium hydroxide solution, (*c*) a solution containing 2 g of hydrochloric acid per dm^3, (*d*) a solution containing 2 g of sodium hydroxide per dm³, (e) 0.1M ethanoic acid (acetic acid), in which the degree of ionization of the acid is 0.014.

(13) Calculate approximate values for the pH of the solutions formed by mixing (a) 25 cm³ of 0.1M NaOH with 50 cm³ of 0.1M HCl, (b) 49 cm³ of 0.1M NaOH with 50 cm³ of 0.1M HCl, (c) 51 cm³ of 0.1M NaOH with 50 cm³ of 0.1M HCl, (d) 100 cm³ of 0.05M HCl with 150 cm³ of 0.04M NaOH. (e) 80 cm³ of 0.1M H_2SO_4 with 120 cm³ of 0.1M KOH.

(14) A 1 M aqueous solution of a weak monobasic acid was found to freeze at $-1.91°C$. Calculate the degree of ionization of the acid and, hence, its dissociation constant and pK_a value. What is the pH of a 0.001M solution of this acid? (K, freezing point constant, is 1.86°C per 1000 g of water.)

(15) The dissociation constant of ethanoic acid (acetic acid) is 1.8×10^{-5} mol dm⁻³. Show that the pH of a solution containing 0.02 of a mole of ethanoic acid and 0.2 of a mole of sodium ethanoate per dm³ is 5.7. Assume the salt to be fully ionized.

(16) Taking the pK_b value of ammonia as 4.699, calculate the change of hydroxide ion concentration produced by adding 10 g of ammonium chloride to 1 dm³ of 0.1M ammonia. Assume the salt to be fully ionized.

(17) Taking the solubility of benzoic acid as 2.44 g dm⁻³ at 0°C in water, and the dissociation constant of the acid as 6×10^{-5} mol dm⁻³, calculate the freezing point at standard pressure of a saturated solution of benzoic acid. (K, freezing point constant, is 1.86°C per 1000 g of water.)

(18) What is the osmotic pressure (in atmospheres) of a 0.1 M solution of ethanoic acid (acetic acid) at 25°C if its dissociation constant at this temperature is 1.8×10^{-5} mol dm⁻³?

(19) Taking the dissociation constant of methanoic acid (formic acid) as 2.14×10^{-4} mol dm⁻³, calculate an approximate value for the degree of ionization of the acid in 0.1M solution, and the concentration in mol dm⁻³ of the acid which has a pH value of 2.398.

SPARINGLY SOLUBLE ELECTROLYTES (SOLUBILITY PRODUCT)

Theory

If a sparingly soluble electrolyte, A_nB_m, dissolves according to the scheme

$$A_nB_m \rightleftharpoons nA^{m+}(aq) + mB^{n-}(aq),$$

its solubility product, K_s, is given by the expression,

$$K_s = [A^{m+}]^n[B^{n-}]^m$$

where, $[A^{m+}]$ and $[B^{n-}]$ are the concentrations of ions in moles per dm^3 of saturated solution of the electrolyte at the stated temperature (usually 25°C).

An important aspect of solubility product is the fact that it defines the point at which the electrolyte is just about to precipitate.

Examples

(1) *If the solubility of barium ethanedioate at room temperature is 0.090 g dm^{-3} of aqueous solution, calculate the solubility product of the salt. What mass of barium ethanedioate is precipitated by adding 2.68 g of sodium ethanedioate (anhydrous) to 1 dm^3 of saturated solution of barium ethanedioate at that temperature?*

The mass of 1 mole of BaC_2O_4 is 225 g, so that 0.090 g of barium ethanedioate is $\frac{0.090}{225}$, or 4×10^{-4} mole. Since the salt dissolves according to the equation

$$BaC_2O_4(s) \rightleftharpoons Ba^{2+}(aq) + C_2O_4^{2-}(aq)$$

the dissolving of 4×10^{-4} mole of salt will produce 4×10^{-4} mole of each of its ions in solution. Therefore,

$$K_s = [Ba^{2+}][C_2O_4^{2-}] = (4 \times 10^{-4})^2 = 1.6 \times 10^{-7} \, mol^2 \, dm^{-6}$$

2.68 g of sodium ethanedioate represents $\frac{2.68}{134}$ mole of the salt, i.e., 2×10^{-2} mole. That is 2×10^{-2} mole of $C_2O_4^{2-}$ are added. Neglecting the relatively small contribution of this ion from barium ethanedioate, we have:

$$[Ba^{2+}][C_2O_4^{2-}] = K_s$$

i.e.,

$$[Ba^{2+}](2 \times 10^{-2}) = 1.6 \times 10^{-7} \, mol^2 \, dm^{-6}$$

$$[Ba^{2+}] = 0.8 \times 10^{-5} \quad or \quad 8 \times 10^{-6} \, mol \, dm^{-3}$$

This means that the concentration of Ba^{2+} is reduced from 4×10^{-4} to $8 \times 10^{-6} \, mol \, dm^{-3}$. Therefore the barium ethanedioate precipitated is

$$4 \times 10^{-4} - 8 \times 10^{-6} = 3.92 \times 10^{-4} \, mole$$

The mass of barium ethanedioate precipitated

$$= 3.92 \times 10^{-4} \times 225 = 8.82 \times 10^{-2} \, g$$

(2) *The solubility product of silver chloride at room temperature is $1 \times 10^{-10} \, mol^2 \, dm^{-6}$. What is the maximum loss from a precipitate of silver* chloride if it is washed with (a) one dm^3 of distilled water, (b) one dm^3 of 0.01M HCl?

(a) The maximum loss is given if the solution used to wash the precipitate becomes saturated with silver chloride at which point the solubility product is just attained in the solution, i.e.,

$$[Ag^+][Cl^-] = 1 \times 10^{-10} \, mol^2 \, dm^{-6}$$

These ionic concentrations are equal, i.e., $[Ag^+] = [Cl^-] = 1 \times 10^{-5}$ mol dm^{-3}. As every mole of AgCl which dissolves produces one mole of Ag^+, the concentration of AgCl in solution (and, therefore, lost from the precipitate) at saturation is 1×10^{-5} mol dm^{-3}. Therefore the mass of silver chloride lost by washing with 1 dm^3 of water $= 1 \times 10^{-5} \times 143.5 = 1.435 \times 10^{-3}$ g.

(b) In 0.01M HCl, assuming complete ionization, $[Cl^-] = 1 \times 10^{-2}$ mol dm^{-3}. With the same reasoning as above, at saturation:

$$[Ag^+][Cl^-] = 1 \times 10^{-10} \, mol^2 \, dm^{-6}$$

The concentration of Cl$^-$ from the silver chloride is negligible compared to that from the HCl, therefore,

$$[Ag^+](1 \times 10^{-2}) = 1 \times 10^{-10} \, mol^2 \, dm^{-6}$$

$$[Ag^+] = 1 \times 10^{-8} \, mol \, dm^{-3}$$

Therefore the mass of silver chloride lost by washing with 1 dm^3 of 0.01M HCl $= 1 \times 10^{-8} \times 143.5 = 1.435 \times 10^{-6}$ g.

Problems on Solubility Product

(Relative atomic masses will be found on page 132)

(20) Saturated solutions of calcium hydroxide were made up in (a) water, and in sodium hydroxide solutions of concentrations, (b) 0.025M, (c) 0.05M, (d) 0.1M. 20 cm^3 of each solution were titrated with 0.05M hydrochloric acid, and required (a) 19.0 cm^3, (b) 23.2 cm^3, (c) 28.4 cm^3, (d) 43.6 cm^3 of the acid. Show that a value of 5.5×10^{-5} mol^3 dm^{-9} can be obtained, from these observations, for the solubility product of calcium hydroxide.

(21) The solubility product of barium carbonate is 7×10^{-9} mol^2 dm^{-6} at 16°C. What is the maximum mass of barium carbonate that can be dissolved in 500 cm^3 of water at this temperature?

(22) The solubility of silver chloride in water at a certain temperature is 1.435 milligrams per dm^3. Calculate the solubility product of silver chloride at this temperature. What mass of silver chloride can be dissolved to make a saturated solution of it in 0.01 M potassium chloride solution at this temperature, assuming both salts to be fully ionized?

(23) At room temperature, the solubility products of silver chloride and silver chromate are respectively 1×10^{-10} mol^2 dm^{-6} and 1×10^{-12} mol^3 dm^{-9}. What is the concentration in mol dm^{-3} of potassium chromate at the end-point when used as indicator in titrating a chloride by silver nitrate?

(24) The solubility product of barium sulphate at room temperature is 1×10^{-10} mol^2 dm^{-6}. Calculate the maximum percentage loss involved in washing at this temperature, a precipitate of barium sulphate of mass 0.500 g with 1 dm^3 of 0.01M sulphuric acid. Assume the acid to be completely ionized.

(25) The solubility product of silver bromide is 4×10^{-13} mol^2 dm^{-6} at 18°C. Calculate the solubility of silver bromide in grams per dm^3 of solution at this temperature. What is the solubility of this salt in potassium bromide solution containing 10 g of potassium bromide per dm^3 at 18°C? Assume both salts to be completely ionized.

(26) The solubility product of calcium sulphate is 6×10^{-5} mol^2 dm^{-6} at room temperature. What mass of calcium sulphate would you expect to be precipitated if 125 cm^3 of 1M sulphuric acid and 125 cm^3 of 1M calcium chloride solution are mixed at room temperature? Assume simple summation of volumes and neglect any possible effect from the hydrochloric acid formed.

(27) The solubility product of silver chloride at the boiling point of water is 21.5×10^{-10} mol^2 dm^{-6}. Would the effect of this salt on the boiling point of water be observable on a Beckmann thermometer reading to one-thousandth of a degree Celsius? ($K = 0.52°$C per 1000 g of water.)

(28) What is the mass of silver in 1 dm^3 of a saturated solution of silver chromate at 25°C if the solubility product of silver chromate at this temperature is 1×10^{-12} mol^3 dm^{-9}? What is the maximum mass of hydrogen chloride that can be dissolved in 250 cm^3 of a saturated solution of silver chromate without causing precipitation of silver chloride? (Solubility product of silver chloride at 25°C is 1×10^{-10} mol^2 dm^{-6}.)

(29) The solubility product of barium ethanedioate at 18°C is 1.2×10^{-7} mol^2 dm^{-6}. Calculate the mass of barium oxalate that would precipitate from 0.5 dm^3 of its saturated solution at 18°C if 0.335 g of anhydrous sodium ethanedioate were dissolved in it without change of volume. Consider the salts completely ionized.

13

Organic Chemistry Formulae of Hydrocarbons by Explosion with Oxygen

Theory

A gaseous hydrocarbon, C_xH_y, explodes with excess oxygen according to the equation:

$$C_xH_y + \left(x + \frac{y}{4}\right)O_2 \rightarrow xCO_2 + \frac{y}{2}H_2O$$

According to Avogadro's Law, the numbers of moles in the balanced equation represent the volumes of reactants and products (if gaseous) taking part in the reaction, i.e., at the same temperature and pressure,

$$1 \text{ vol. of } C_xH_y + \left(x + \frac{y}{4}\right) \text{ vol. of } O_2$$
$$\rightarrow x \text{ vol. of } CO_2 + \text{negligible vol. of } H_2O \text{ as water}$$

The volume relations are as shown, the water formed occupying negligible volume at room temperature and pressure. The volumes of the hydrocarbon, and of the oxygen added, are noted. After the explosion, carbon dioxide is absorbed in KOH solution and the reduction in volume is noted. The only residual gas is then unused oxygen. All readings must be at constant room temperature and pressure.

It follows from the above equation that:

Vol. of carbon dioxide produced = x × (volume of hydrocarbon used)

Vol. of oxygen used up by the reaction = $\left(x + \frac{y}{4}\right)$ × (volume of hydrocarbon used)

These equations can be solved for x and y.

Sometimes it is more convenient to use the initial contraction in volume as one of the simultaneous equations involving x and y, i.e.,

Initial contraction in vol. = vol. of hydrocarbon used
+ vol. of oxygen used
− (vol. of carbon dioxide produced)
= (vol of hydrocarbon)
+ $\left(x + \dfrac{y}{4}\right)$ × (vol. of hydrocarbon)
− x × (vol. of hydrocarbon)

Example

60 cm^3 of oxygen were added to 10 cm^3 of a gaseous hydrocarbon. After explosion and cooling, the gases occupied 50 cm^3 and, after absorption by KOH solution, 30 cm^3 of oxygen remained. Calculate the molecular formula of the hydrocarbon. (*Temperature and pressure constant at room values.*)

$$C_xH_y + \left(x + \dfrac{y}{4}\right)O_2 \rightarrow xCO_2 + \dfrac{y}{2}H_2O$$

From this equation,

Vol. of carbon dioxide produced
= x (volume of hydrocarbon used) . . . . (i)

Vol. of oxygen used up
= $\left(x + \dfrac{y}{4}\right)$ (volume of hydrocarbon used) . . . (ii)

In the above data,

Volume of carbon dioxide = (50 − 30) cm^3 = 20 cm^3
Volume of oxygen used up = (60 − 30) cm^3 = 30 cm^3
Volume of hydrocarbon = 10 cm^3

From (i), $20 = x \times 10$, i.e., $x = 2$

From (ii) $30 = \left(2 + \dfrac{y}{4}\right)(10)$, i.e., $y = 4$

The hydrocarbon is C_2H_4, ethene (ethylene).

Problems on Formulae of Hydrocarbons by Explosion with Oxygen

(*Relative atomic masses will be found on page* 132)

(1) To 20 cm^3 of a gaseous hydrocarbon, 80 cm^3 of oxygen were added. After explosion and cooling to room temperature, the residual gases occupied 70 cm^3. After absorption by KOH, 30 cm^3 of oxygen remained. Determine the formula of the hydrocarbon. Measurements made at the same temperature and pressure throughout.

(2) 25 cm^3 of a gaseous hydrocarbon were mixed with 200 cm^3 of oxygen and exploded. After cooling the residual gases occupied 137.5 cm^3. The volume was reduced by 100 cm^3 on exposure of the gases to potassium hydroxide solution, and the remaining gas was oxygen. What is the formula of the hydrocarbon?

(3) To 30 cm^3 of a gaseous hydrocarbon, 150 cm^3 of oxygen were added. After explosion and cooling to room temperature, 105 cm^3 of gas remained. By absorption with KOH solution a diminution of 60 cm^3 was produced, and the remaining gas was shown to be oxygen. Determine the formula of the hydrocarbon. (Pressure constant at atmospheric.)

(4) 90 cm^3 of oxygen were added to 15 cm^3 of a hydrocarbon. After explosion and cooling, the residual gases occupied a volume of 75 cm^3, and, after absorption by KOH solution, the residual oxygen occupied 45 cm^3. Find the formula of the hydrocarbon. (Temperature and pressure constant.)

(5) On exploding 30 cm^3 of a gaseous hydrocarbon with 100 cm^3 of oxygen and cooling, the residual gases occupied a volume of 70 cm^3 and, after absorption of the carbon dioxide, the residual gas occupied a volume of 10 cm^3 and was entirely oxygen. What is the formula of the hydrocarbon? (All measurements were made at room temperature and pressure.)

(6) 11.5 cm^3 of a gaseous hydrocarbon were exploded with excess oxygen and there was a diminution in volume of 34.5 cm^3 and on absorption with KOH solution there was a further reduction in volume of 34.5 cm^3. What is the molecular formula of the hydrocarbon? (All measurements were made at room temperature and pressure.)

14

Formulae of Organic Compounds

Theory

The common elements found in organic compounds are carbon, hydrogen, oxygen, nitrogen, chlorine, bromine and sulphur, together with certain metals. The following account summarizes the methods of estimating the non-metallic elements.

(1) CARBON AND HYDROGEN

The compound is oxidized by heating in oxygen (with precautions to ensure complete oxidation). Hydrogen is oxidized to water and weighed after absorption in concentrated sulphuric acid. Carbon is oxidized to carbon dioxide and weighed after absorption in KOH bulbs or in soda-lime. From $M_r(CO_2) = (12 + 32) = 44$ and $M_r(H_2O) = (2 + 16) = 18$, if a g of a compound produce c g of carbon dioxide and w g of water,

$$\text{Percentage of carbon} = \frac{12}{44} \times c \times \frac{100}{a}$$

$$\text{Percentage of hydrogen} = \frac{2}{18} \times w \times \frac{100}{a}$$

(2) NITROGEN

(*a*) The compound is combusted in such a way that the nitrogen is liberated and measured as gas. From the relation:

$$N_2$$

1 mole is 28 g   or   22 400 cm³ at s.t.p.

if a g of a compound produced n cm³ of nitrogen converted to s.t.p.,

$$\text{Percentage of nitrogen} = \frac{n}{22\,400} \times 28 \times \frac{100}{a}$$

(b) The nitrogen is converted to ammonium sulphate by heating with concentrated sulphuric acid and potassium hydrogensulphate. After dilution, the ammonia is expelled by heating with excess of NaOH solution and estimated by absorption in standard acid.

$$N \equiv NH_3 \equiv H^+$$
$$14\,g \qquad 1000\,cm^3 \; 1M\; HCl$$

If a g of a compound produce ammonia equivalent to x cm^3 of 1M HCl.

$$\text{Percentage of nitrogen} = \frac{x}{1000} \times 14 \times \frac{100}{a}$$

(3) CHLORINE AND BROMINE

These halogens are converted to the corresponding silver compound by heating with silver nitrate and fuming nitric acid in a sealed tube. The silver halide is weighed after purification. From the relations:

$$M_r(AgCl) = (108 + 35.5) = 143.5$$
and
$$M_r(AgBr) = (108 + 80) = 188$$

if a g of a compound produce b g of either AgCl or AgBr,

$$\text{Percentage of chlorine} = b \times \frac{35.5}{143.5} \times \frac{100}{a}$$

or
$$\text{Percentage of bromine} = b \times \frac{80}{188} \times \frac{100}{a}$$

(4) SULPHUR

Sulphur is converted to sulphuric acid by heating with fuming nitric acid in a sealed tube and is then precipitated and weighed as barium sulphate. From the relation:

$$M_r(BaSO_4) = (137 + 32 + 64) = 233$$

if a g of a compound produce w g of barium sulphate,

$$\text{Percentage of sulphur} = w \times \frac{32}{233} \times \frac{100}{a}$$

(5) OXYGEN

This element is usually estimated as the difference from 100% after accounting for all other elements.

When the percentages of all elements are known, the *empirical formula* of the compound can be calculated. The following example is typical.

A compound contained the following percentages of the named elements:

	Carbon	Hydrogen	Nitrogen	Oxygen
	42.9	2.4	16.7	38.1

Divide each by the relative atomic mass.

No. of moles of atoms are proportional to:

$$\frac{42.9}{12} \qquad \frac{2.4}{1} \qquad \frac{16.7}{14} \qquad \frac{38.1}{16}$$

$$= 3.58 \qquad = 2.4 \qquad = 1.19 \qquad = 2.38$$

Divide each by the smallest.

The atomic ratio is:

$$\frac{3.58}{1.19} \qquad \frac{2.4}{1.19} \qquad \frac{1.19}{1.19} \qquad \frac{2.38}{1.19}$$

$$= 3 \qquad = 2 \qquad = 1 \qquad = 2$$

The empirical formula is $C_3H_2NO_2$. (The calculation may not give actual integers at this stage. It may give ratios such as 1.5:1, or 1.33:1 or 1.67:1. These are rendered in the lowest integral equivalents, i.e., 3:2, 4:3 and 5:3 respectively.)

To find the molecular formula of the compound, the relative molecular mass of the compound must be known. In this case, it is about 170, determined by freezing point depression in a suitable solvent, probably tribromomethane. From this,

$$M_r(C_3H_2NO_2)_n = 170$$

i.e., $\qquad 84n = 170$

or $\qquad n = 2$

That is, the molecular formula is $C_6H_4N_2O_4$.

To find the structural formula, suitable reactions of the compound must be known and used to deduce the groupings present.

Problems on Formulae of Organic Compounds from Various Data

(Relative atomic masses will be found on page 132)

(1) An organic acid was shown on analysis to contain C 40%, H 6.7%, O 53.3%. If the relative molecular mass of the acid was 60, identify the acid.

(2) A compound containing only C, H and O contains C 52.2%, H 13.1%. The relative vapour density of the compound is 23. Calculate its molecular formula.

(3) A compound of carbon with hydrogen contains 7.7% of hydrogen. The relative vapour density of the compound is 39. Calculate the molecular formula.

(4) 0.493 g of a compound containing carbon, hydrogen and oxygen gave 0.881 g of CO_2 and 0.359 g of H_2O on combustion. 0.082 g of the compound displaced 27.0 cm^3 of air, in a Victor Meyer determination, measured at 17°C and 98 650 N m^{-2} (750 mmHg). Calculate the molecular formula of the compound.

(5) An aromatic hydrocarbon contained 91.3% of carbon. Suggest a formula for the compound and one way in which it might have been made from benzene.

(6) 0.5 g of an organic compound containing carbon, hydrogen and oxygen gave an analysis 0.6875 g of CO_2 and 0.5625 g of H_2O. Find the empirical formula of the substance. If it was of relative vapour density 16, suggest a structural formula for it.

(7) 0.620 g of an organic compound gave on combustion 1.760 g CO_2 and 0.420 g H_2O. 0.232 g of the same compound gave 29.5 cm^3 of nitrogen measured dry at 15°C and 101 300 N m^{-2} (760 mmHg). 0.83 g of the compound dissolved in 134 g of benzene gave an elevation of 0.178°C in the boiling point. ($K = 2.67$°C per 1000 g of benzene.) Calculate the molecular formula of the compound.

(8) A compound consisted of C 62%, H 10.35%, O 27.6%. It formed an oxime with hydroxylamine but did not reduce Fehling's solution. The relative vapour density was 29. Write the structural formula for the compound and give the equations for the reactions it would undergo with the above-mentioned reagents.

(9) Determine the structural formula of an organic compound from the following data:

0.708 g of the compound gave on combustion 1.056 g of CO_2 and 0.540 g H_2O.

By Kjeldahl's method 1.23 g of the compound yielded ammonia which required 20.8 cm^3 of 0.5M sulphuric acid for neutralization.

The compound gave off ammonia on boiling with sodium hydroxide solution, and the dried sodium compound so produced yielded methane on heating with soda lime.

(10) An aliphatic primary amine was converted into the hexachloroplatinate(IV) and 1 g of this left a residue of 0.390 g of platinum. Identify the amine.

(11) 0.60 g of the silver salt of a monobasic organic acid gave a residue of 0.283 g of silver on ignition. The acid contained 68.8% of carbon. Suggest a molecular formula for the acid.

(12) 1.200 g of a compound containing carbon, hydrogen and oxygen gave on combustion 1.173 g CO_2 and 0.240 g H_2O.
1.125 g of the compound in 125 g of water gave a solution freezing at $-0.186°C$. Calculate the relative molecular mass of the compound and write its molecular formula.
($K = 1.86°C$ per 1000 g of water)

(13) 0.72 g of a compound gave on combustion 1.615 g CO_2 and 0.99 g H_2O. 0.42 g of the same compound gave 84 cm³ of nitrogen by Dumas' method. (Measured dry at 15°C and 101 300 N m^{-2} (760 mmHg).) It reacted with nitrous acid to form a salt and gave a crystalline compound with iodomethane. Write down the structural formula for the compound and the equations for the action of nitrous acid on its isomers.

(14) 1.0 g of a compound which contained only carbon, hydrogen and oxygen gave on analysis 1.695 g CO_2 and 1.075 g H_2O. The substance reduced diamminesilver(I) nitrate, being oxidized at the same time to an acid. Write down the name and structural formula for the compound.

(15) An aromatic organic compound containing only carbon, hydrogen and nitrogen contained C 79.3%, H 9.07%. It was basic in character and formed a nitroso compound with nitrous acid, the nitroso group being attached to the benzene nucleus. Identify the compound and write its structural formula. What would be its action on iodomethane?

(16) An organic compound gave on oxidation an acid having the same number of carbon atoms in the molecule and with the following composition: C 40%, H 6.67%, and the rest oxygen. The original substance reduced Fehling's solution and formed an addition compound with ammonia. Write the structural formula for the original compound.

(17) An organic compound contained 65.4% of carbon and 9.09% of hydrogen and the rest was nitrogen. On reduction it gave a compound having a relative molecular mass 4 greater than that of the original substance. Write a structural formula for the original substance and also for the reduction product. Name them.

(18) 1.363 g of an organic compound gave on combustion 1.100 g of CO_2 and 0.563 g of H_2O. Also 1.453 g of the compound gave, by a Carius determination, 2.507 g of silver bromide. The relative vapour density of the compound was found to be about 55. Find its molecular formula.

(19) An aromatic organic compound containing carbon, hydrogen and chlorine contained 5.534% of hydrogen and 1.00 g of the compound gave 1.134 g of silver chloride. The original substance gave an alcohol on hydrolysis. Give its structural formula and that of one of its isomers.

(20) Two organic compounds both gave the following figures on analysis:

  1.08 g of the compound gave on combustion 0.960 g CO_2 and 0.393 g H_2O.

  0.90 g of the compound gave 2.61 g of AgCl by a Carius determination.

The relative vapour density of each compound is about 50. Write formulae for the two compounds and state briefly how you would proceed to obtain the experimental evidence necessary to assign the correct structure to each.

(21) On analysis, a compound was found to contain C = 42.9%, H = 2.4%, N = 16.7%, O = 38.1%. On reduction it gave a diamine, which, with nitrous acid, gave a yellow-brown dye. Write down a structural formula for the original compound. Its relative molecular mass was found to be about 170.

(22) Two organic compounds A and B have the same percentage composition, C = 52.2%, H = 13.1%, O = 34.7%. The relative vapour density of both is 23. A yielded hydrogen when acted upon by sodium and gave steamy fumes of hydrogen chloride with phosphorus pentachloride. B gave no visible effect with either test. Assign structural formulae to A and B.

(23) Two isomeric organic compounds contained 78.5% of carbon, 8.42% of hydrogen and 13.08% of nitrogen. Both gave a vapour possessing a very disagreeable odour on being boiled with alcoholic KOH solution and trichloromethane. With nitrous acid one gave a diazonium compound but the other gave an alcohol with no intermediate diazonium compound. Suggest structural formulae for the two compounds.

(24) 0.8 g of an organic compound, consisting of carbon, hydrogen and oxygen contained 0.533 g of carbon and 0.0889 g of hydrogen. The compound gave an oxime with hydroxylamine and formed a hydrazone with hydrazine. On oxidation it formed an acid possessing a smaller number of carbon atoms and also carbon dioxide. Write down the name and formula of the original compound.

(25) Several organic compounds give the following figures when analysed:

  0.649 g of the compound on combustion gave 1.450 g CO_2 and 0.890 g H_2O.

  0.147 g of the compound gave 29.8 cm^3 of nitrogen measured at 15°C and 100 000 N m^{-2} (750 mmHg).

The relative vapour density of the compounds is about 30. Find the common molecular formula, and write the structural formulae for all possible isomers.

(26) On acting upon a simple ketone with phosphorus pentachloride, a compound was formed, 0.61 g of which gave 1.55 g of silver chloride by Carius' method. Name the halogen compound and write its structural formula.

(27) Two organic compounds A and B gave the following figures on analysis: 0.870 g of one compound on combustion gave 1.980 g CO_2 and 0.810 g H_2O. In a Victor Meyer determination, 0.058 g of the compound displaced 24.2 cm^3 of air measured at 15°C and 98 650 N m^{-2} (740 mmHg). Both compounds give an oxime with hydroxylamine. A reduces diamminesilver(I) nitrate but B does not. Assign a structural formula to each.

(28) 1.00 g of an organic compound of relative vapour density 59.75 gave 3.602 g of silver chloride. The compound contains 10.05% of carbon. Identify it.

(29) 0.500 g of an organic compound which contained carbon, hydrogen and oxygen gave on combustion 1.428 g CO_2 and 0.333 g H_2O. On treating the compound with the phosphorus pentachloride a very reactive monohalogen compound was formed and hydrogen chloride liberated. Assign a structural formula to the original compound.

(30) A hydrocarbon contained 83.3% of carbon, and had a relative vapour density of 36. Write down its molecular formula and the structural formulae of its various isomers and name them.

(31) Two organic compounds, A and B, containing carbon, hydrogen and oxygen have identical molecular formulae. 0.80 g of A gave on combustion 1.76 g CO_2 and 0.96 g H_2O. The relative vapour density of B was found to be about 30. On vigorous oxidation A gave an acid containing C 48.7%, H 8.1%, O 43.2%, whilst B evolved carbon dioxide and left an acid containing C 40.0%, H 6.7%, O 53.3%. Assign structural formulae to A and B.

(32) A halogen compound contained 54.47% of chlorine and 0.200 g of the compound gave on combustion 0.3152 g CO_2 and 0.0460 g H_2O. On hydrolysis it gave a monobasic aromatic acid which contained no halogen. Identify the compound.

(33) 1.00 g of an organic compound containing carbon, hydrogen and oxygen gave on analysis 2.81 g of CO_2 and 0.574 g H_2O. It gave a purple colour with iron(III) chloride and dissolved in sodium hydroxide. Identify the compound.

(34) 0.40 g of an organic compound containing C, H and O gave on analysis 0.88 g of CO_2 and 0.48 g H_2O. By treatment with hydrogen iodide a mixture of iodomethane and iodoethane was formed. Write down the structural formula of the substance and of two isomers.

(35) On analysis an organic compound was found to contain 79.3% of carbon, 5.66% of hydrogen, and the rest was oxygen. On oxidation

it formed a monobasic acid, the silver salt of which was found to contain 47.1% of silver. Identify the original compound.

(36) An organic compound containing C, H, O and S was found to contain 31.4% of carbon and 2.52% of hydrogen. 0.70 g of it was found to yield, by Carius' method, 1.37 g of barium sulphate. On hydrolysis a hydrocarbon was formed of relative vapour density 39. Suggest a structural formula for the compound.

(37) An aromatic compound gave the following results on analysis:

0.920 g gave on combustion 1.760 g CO_2 and 3.60 g H_2O.

0.138 g gave 23.6 cm^3 of nitrogen measured dry at 15°C and 101 600 N m^{-2} (762 mmHg).

0.828 g lowered the freezing point of 120 g of benzene by 0.25°C. ($K = 5.0$°C per 1000 g of benzene.)

The compound is a yellow solid and is soluble in hydrochloric acid with the formation of a chloride. Write formulae for the three isomers which satisfy the given conditions.

(38) 1 mole of a pentapeptide gave the following products on complete hydrolysis: 2 moles of alanine (A), 1 mole of glycine (G), 1 mole of serine (S), and 1 mole of leucine (L).

Two of the products identified after partial hydrolysis were the dipeptides glycylalanine (G.A.) and leucylalanine (L.A.). Partial hydrolysis under different conditions gave, amongst other products, the dipeptide serylleucine (S.L.) and the tripeptide alanylserylleucine (A.S.L.). From these results determine the sequence of amino acids in the pentapeptide.

15

Gas Analysis

Theory

Problems on gas analysis can vary considerably in type. They are usually solved by writing an equation for the behaviour of each gas during the analysis and deriving, from these chemical equations, corresponding algebraic equations, which can be solved. The following examples are typical.

Examples

(1) *24 cm³ of a mixture of methane and ethane were exploded with 90 cm³ of oxygen. After cooling to room temperature, the volume of gas was noted. It was found to decrease by 32 cm³ when treated with KOH solution (pressure constant throughout). Calculate the composition of the mixture.*

Let there be a cm³ of methane and b cm³ of ethane, so that

$$a + b = 24 \quad \ldots \quad (i)$$

The equations and, by applying Avogadro's Law, the volumes involved are:

$CH_4 \;+\; 2O_2 \;\to\; CO_2 \;+\; 2H_2O$
1 vol   1 vol   —   (Volume of H_2O as liquid, after
a cm³ a cm³   cooling, is negligible)

$C_2H_6 \;+\; 3\tfrac{1}{2}O_2 \;\to\; 2CO_2 \;+\; 3H_2O$
1 vol   2 vol   —
b cm³ $2b$ cm³

The carbon dioxide produced is absorbed by the KOH solution and its volume is, therefore, 32 cm³. From this:

$a + 2b = 32$

From (i) above, $\quad a + b = 24$

Subtracting, $\quad b = 8$

That is, there are 8 cm³ of ethane and 16 cm³ of methane in the mixture.

(2) 32 cm³ of a mixture of carbon monoxide, methane and hydrogen were mixed with 50 cm³ of oxygen and exploded. After cooling to room temperature, the volume was noted. It was reduced by 22 cm³ when exposed to KOH solution, leaving 16 cm³ of excess oxygen. Calculate the composition of the mixture.

Let there be a cm³ of carbon monoxide, b cm³ of methane and c cm³ of hydrogen. Then,

$$a + b + c = 32 \qquad \qquad \text{(i)}$$

The equation and volume relations involved are:

$$2CO + O_2 \rightarrow 2CO_2$$
2 vol   1 vol   2 vol

a cm³ $\frac{a}{2}$ cm³ a cm³

$$CH_4 + 2O_2 \rightarrow CO_2 + 2H_2O$$
1 vol   2 vol   1 vol   ——   (See note, p. 111)

b cm³ $2b$ cm³ b cm³

$$2H_2 + O_2 \rightarrow 2H_2O$$
2 vol   1 vol   ——

c cm³ $\frac{c}{2}$ cm³

The volume of carbon dioxide is the reduction produced by KOH solution, i.e., 22 cm³. From this:

$$a + b = 22 \qquad \qquad \text{(ii)}$$

From (i) $a + b + c = 32$

i.e., $c = 10$

The volume of oxygen used is $(50 - 16)$ cm³, i.e., 34 cm³, so that

$$\frac{a}{2} + 2b + \frac{c}{2} = 34$$

Since $c = 10$,

$$\frac{a}{2} + 2b = 29$$

From (ii), $2a + 2b = 44$

Subtracting, $1\tfrac{1}{2}a = 15$

i.e., $a = 10$, so $b = 12$

The mixture had the composition, methane, 12 cm³, carbon monoxide 10 cm³ and hydrogen 10 cm³.

Problems on Gas Analysis

(Relative atomic masses will be found on page 132)

(1) 45 cm^3 of air freed from carbon dioxide were mixed with hydrogen, the total volume being 100 cm^3. The mixture was exploded and cooled and the residual gas occupied 71.6 cm^3. Calculate the percentage of oxygen in the sample of air. (Temperature and pressure constant at the ordinary atmospheric values.)

(2) 75 cm^3 of a mixture containing 30% methane and 70% hydrogen by volume were mixed at room temperature with 200 cm^3 of oxygen and exploded. What is the residual volume on cooling again to room temperature? (Pressure constant at 760 mmHg).

(3) 65 cm^3 of a mixture of hydrogen, carbon monoxide and nitrogen are mixed with 50 cm^3 of oxygen at room temperature. The mixture is exploded and allowed to cool. The residual volume is 55 cm^3. After absorption with KOH solution the final volume is 25 cm^3. Calculate the percentage by volume of each gas in the original mixture. (Temperature and pressure constant at room values.)

(4) 57 cm^3 of a mixture of methane and oxygen (in which the ratio of the volume of oxygen to methane was greater than 2:1) were exploded and the volume of residual gas was 23 cm^3. What was the volume of methane in the original mixture? (All measurements made at 15°C and 760 mmHg.)

(5) 60 cm^3 of a mixture of methane, ethane and oxygen were exploded and the volume of residual gas when cooled was 27.5 cm^3. After absorption with KOH solution the volume fell to 7.5 cm^3, the final residue being oxygen. Calculate the composition of the mixture. (All measurements made at 15°C and 760 mmHg.)

(6) 30 cm^3 of a mixture of methane and ethane were mixed with 100 cm^3 of oxygen at ordinary temperature and pressure and exploded. After cooling the residual gas occupied 61.5 cm^3. Find the percentage by volume of each gas in the mixture.

16

Miscellaneous Problems

The following problems cover some kinds of calculation which have not been studied before in this book.

Theory

SOLUBILITY

The solubility of a compound or element in water is defined as the number of grams of the compound or element required to saturate 1000 grams of water at the temperature under consideration.

Solubility is determined, in general, by preparing a saturated solution of the material at the relevant temperature and then analysing it. Analysis may be by evaporation or by chemical analysis.

Examples

(1) *A saturated solution (at 20°C) of a compound, X, was analysed by evaporation to dryness with the following results:*

  Mass of evaporating dish   = 20.110 g
  Mass of evaporating dish and solution = 43.989 g
  Mass of evaporating dish and solid X = 23.991 g

Calculate the solubility of X *at 20°C.*

From the figures given,

  Mass of solid X = 3.881 g
  Mass of water present = 19.998 g

Therefore,

  Solubility of $X = \dfrac{3.881}{19.998} \times 1000 = 194.0$ g per 1000 g of water at 20°C.

(2) 5.372 g *of a saturated solution of common salt at* 20°C *were made up to* 250 cm³ *with distilled water.* 25.0 cm³ *of this diluted solution required* 22.53 cm³ *of* 0.1025M *silver nitrate solution for titration. Calculate the solubility of common salt at this temperature.*

The titration equation is:

$$Ag^+ + Cl^- \rightarrow AgCl$$

or

$$NaCl + AgNO_3 \rightarrow AgCl + NaNO_3$$

From the equation,

1 mole of $AgNO_3$ reacts with 1 mole of NaCl, therefore,

1000 cm³ of 1M $AgNO_3$ reacts with 58.5 g of NaCl

22.53 cm³ of 0.1025M $AgNO_3$ reacts with

$$58.5 \times \frac{22.53}{1000} \times 0.1025 \text{ g of NaCl}$$

This is the mass of NaCl in 25 cm³ of the diluted solution, therefore, the mass of NaCl in 250 cm³ of the diluted solution

$$= 58.5 \times \frac{22.53}{1000} \times 0.1025 \times 10 = 1.351 \text{ g}$$

Therefore, the mass of water in the standard solution

$$= (5.372 - 1.351) \text{ g} = 4.021 \text{ g}$$

Therefore, the solubility of the common salt

$$= \frac{1.351}{4.021} \times 1000 = 336.0 \text{ g per 1000 g of water at } 20°C$$

Theory

STEAM DISTILLATION

If an immiscible liquid, X, is distilled with water, each liquid makes its own contribution to the vapour pressure. The mixture boils when the sum of the vapour pressures is equal to the atmospheric pressure. If, at this point, the vapour pressure of X is x, and of water is y, so that $(x + y)$ is the atmospheric pressure, the following relationships hold:

$$\frac{x}{y} = \frac{\text{Number of moles of X in the vapour (and hence in the distillate)}}{\text{Number of moles of water in the vapour (and hence in the distillate)}}$$

but the number of moles of X = $\dfrac{\text{mass of X in vapour}}{\text{molar mass of X}}$

and number of moles of steam = $\dfrac{\text{mass of steam}}{\text{molar mass of steam}}$

therefore,

$$\dfrac{x}{y} = \dfrac{\text{mass of X in vapour} \times \text{molar mass of steam}}{\text{mass of steam} \times \text{molar mass of X}}$$

Examples

(1) *When a compound X of relative molecular mass 120 is steam distilled, the mixture boils at 99.3°C, at which temperature the vapour pressure of water is $98\,650\,N\,m^{-2}$ (740 mmHg). If the atmospheric pressure is $102\,900\,N\,m^{-2}$ (765 mmHg), calculate the composition of the distillate by mass.*

Vapour pressure of $X = (102\,900 - 98\,650) = 4250\,N\,m^{-2}$. From the data,

$$\dfrac{\text{Mass of vapour of } X}{\text{Mass of steam}} = \dfrac{120 \times 4250}{18 \times 98\,650} = \dfrac{1}{4.44}$$

Therefore, the percentage by mass of X in the distillate

$$= \dfrac{1}{5.44} \times 100 = 18.4$$

(2) *When a compound X is steam distilled, the liquid boils when the vapour pressure of water is $96\,000\,N\,m^{-2}$ (720 mmHg) and the atmospheric pressure is $102\,900\,N\,m^{-2}$ (765 mmHg). If the distillate contains 26% of X by mass, calculate the relative molecular mass of X.*

Let the relative molecular mass of X be M. The vapour pressure of X is $(102\,900 - 96\,000)\,N\,m^{-2}$ at the boiling point of the mixture, i.e., $6900\,N\,m^{-2}$. From the data,

$$\dfrac{6900}{96\,000} = \dfrac{26}{74} \times \dfrac{18}{M}$$

Therefore,

$$M = \dfrac{26 \times 96\,000 \times 18}{74 \times 6900} = 101$$

That is the relative molecular mass of the compound is 101.

Miscellaneous Problems

(*Relative atomic masses will be found on page* 132)

(1) 1 dm^3 of a mixture of oxygen and ozone, measured at s.t.p., was allowed to remain in contact with 100 cm^3 of acidified potassium iodide solution until the reaction was complete. 25 cm^3 of the iodide solution required 32.3 cm^3 of 0.1M sodium thiosulphate to react with the iodine liberated. Calculate the percentage by volume of ozone (trioxygen, O_3) in the mixture.

(2) A mixture known to contain only the radicals SO_4^{2-}, Br^-, K^+ and NH_4^+ was analysed in the following way:

(*a*) 2.008 g of the mixture gave 1.864 g of barium sulphate.

(*b*) 3.012 g of the mixture were made up to 100 cm^3 of solution and 25 cm^3 of this required 30.0 cm^3 of 0.1M $AgNO_3$.

(*c*) 2.259 g of the mixture left, on ignition with concentrated sulphuric acid to constant mass, 0.783 g of potassium sulphate.

Estimate the percentage by mass of each radical. From what salts was the mixture made up?

(3) Show by means of equations what you consider to be the effect of varying degrees of heat on copper(II) sulphate crystals from the following results:

Mass of crucible and lid	9.0400 g
Mass of crucible and lid and crystals . .	10.2100 g
Mass after heating to constant mass at 100°C .	9.8724 g
Mass after heating to constant mass at 230°C .	9.7880 g
Mass after heating very strongly . . .	9.4128 g

(4) 100 cm^3 of a sulphurous acid solution required 11.2 cm^3 0.05M iodine solution for complete reaction. Calculate the volume of dissolved sulphur dioxide at 15°C and 98 920 N m^{-2} (742 mmHg) per dm^3 of the acid.

(5) Water from a certain spring gave, on boiling, a gaseous mixture containing 21.0% oxygen, 43.6% nitrogen and the remainder carbon dioxide (by volume). Calculate the percentages by volume of the mixture of gases with which the water had been in contact, assuming s.t.p. conditions. The absorption coefficients for gases are at s.t.p. oxygen 0.04, nitrogen 0.02, carbon dioxide 1.79.

(6) A dish was weighed and weighed again together with a piece of dry marble. Water was added and then exactly 10 cm^3 of concentrated hydrochloric acid and a watch glass placed over the dish to prevent loss. When effervescence had ceased, the marble was washed and dried

in the dish in a steam oven. Calculate the percentage of hydrogen chloride in the specimen of the acid from the following readings:

Mass of dish . . . . 26.320 g
Mass of dish and marble . . . . . 27.760 g
Mass of dish and marble after expt. . . . 27.198 g
Density of the acid = $1.18\,\text{g cm}^{-3}$.

(7) 0.290 g of an ester formed from a monobasic acid and a monoacidic base were boiled for some time with $100\,\text{cm}^3$ of 0.1M KOH solution. The resulting solution required $60.8\,\text{cm}^3$ 0.1M HCl for neutralization. What was the relative molecular mass of the ester?

(8) $100\,\text{cm}^3$ of a sample of coal gas, containing 50% hydrogen, 34% methane, 8% carbon monoxide, 4% ethene and 4% nitrogen by volume, were mixed with $180\,\text{cm}^3$ of oxygen and exploded. Calculate the percentage composition by volume of the residual gas on cooling to the original room temperature. (Assume presure constant and water formed entirely liquefied.)

(9) 10.2 g of ammonium iron(II) sulphate were made up to $250\,\text{cm}^3$ with acidified water. 1.07 g of a mixture of potassium manganate(VII) and potassium sulphate was made up to $250\,\text{cm}^3$ and $37.5\,\text{cm}^3$ of this solution were required to react with $25\,\text{cm}^3$ of the ammonium iron(II) sulphate solution. Calculate the percentage of potassium manganate(VII) in the mixture.

(10) What will be the osmotic pressure at $0°C$ of a solution of 10.5 g of mannitol per dm^3? What would be the concentration of a solution of urea (carbamide) which would be isotonic with it?

Mannitol, $C_6H_{14}O_6$; Urea, CON_2H_4

(11) The carbon dioxide in the air of a room was estimated by drawing $200\,\text{dm}^3$ of it at $14°C$ and $100\,400\,\text{N m}^{-2}$ (753 mmHg) through potassium hydroxide bulbs of known mass. The increase in mass of the bulbs was 0.157 g. Calculate the percentage by volume of carbon dioxide in the air.

(12) 1 g of a specimen of sodium peroxide was dissolved in water and the solution boiled for some time. It was then made up to $250\,\text{cm}^3$ with distilled water, and $25\,\text{cm}^3$ of this required $24.7\,\text{cm}^3$ 0.1M HCl for neutralization. Calculate the percentage purity of the specimen.

(13) The solubility of a substance X is five times as great in liquid A as it is in liquid B, and A and B are immiscible. Compare the quantities of X extracted from $1\,\text{dm}^3$ of B by

(a) $1\,\text{dm}^3$ of A.
(b) Two successive washings and separations with $500\,\text{cm}^3$ of A.

(14) How many kg of iron pyrites, FeS_2, would be required to produce 100 kg of sulphuric acid containing 98% H_2SO_4? How many m³ of air containing 21% of oxygen by volume and measured at 15°C and 100 000 N m⁻² (750 mmHg), would be required for complete combustion of the pyrites?

(15) Calculate the solubility of sodium chloride per 1000 g of water at 20°C from the following data:

21.5 g of saturated solution of common salt at 20°C were made up to 1 dm³ with distilled water. 25 cm³ of this solution required 23.2 cm³ of 0.1M $AgNO_3$.

(16) A sample of water-gas contained 45% hydrogen, 0.5% methane, 44% carbon monoxide (by volume) and the remainder incombustible gases. Calculate the percentage composition by volume of the gas left after 100 cm³ of the water-gas have been exploded with 100 cm³ of oxygen and the mixture cooled to ordinary temperature. (Assume pressure constant and all water condensed as liquid.)

(17) Find the percentage of PO_4^{3-} ion in disodium hydrogenphosphate(V) crystals from the following data:

1.320 g of the crystals was dissolved in water and ammonium chloride and ammonia added, and the solution boiled. Excess magnesium sulphate solution was added, and the precipitate formed was filtered off, washed, ignited and weighed. The residue, which was magnesium heptaoxodiphosphate(V), $Mg_2P_2O_7$, had a mass of 0.8187 g. Hence find the number of moles of water of crystallization in 1 mole of the original salt.

(18) 1.843 g of an organic compound containing C, H, O and S gave 3.08 g CO_2 and 0.63 g of H_2O, whilst 1.327 g of the same compound gave 1.942 g $BaSO_4$. The relative molecular mass of the compound was about 160.

Write a formula for the substance and suggest by equations how it would react on,

(a) hydrolysis with concentrated hydrochloric acid,

(b) treatment with sodium hydroxide,

(c) treatment with potassium cyanide.

(19) A current of 3 amperes was passed for 35 minutes through solutions of copper(II) sulphate, silver nitrate and acidified water. If the mass of copper deposited was 2.069 g, calculate (a) the electrochemical equivalent of copper, (b) the mass of silver deposited, (c) the volume of hydrogen measured dry at 15°C and 102 700 N m⁻² (770 mmHg) pressure.

(20) Calculate the solubility of chlorine in grams of gas per 1000 g of water at 15°C from the following data:

From a bottle with a mass of 150.3 g, saturated chlorine water at 15°C was rapidly poured into excess of potassium iodide solution. The bottle and residual chlorine water had a mass of 49.5 g. The liberated iodine was made up to 250 cm^3 and 25 cm^3 of the solution required 20.0 cm^3 of 0.1M sodium thiosulphate.

(21) Calculate the freezing point of a solution of urea (CON_2H_4) which contains 1 g urea in 100 g of water from the following data:

The freezing point of a solution of 7.3 g of cane sugar in 79.0 g of water was −0.500°C.

Cane sugar is $C_{12}H_{22}O_{11}$.

(22) 1 g of impure calcium carbonate (containing insoluble earthy matter) was dissolved in dilute hydrochloric acid, and the calcium precipitated as ethanedioate by the addition of ammonia and ammonium ethanedioate. The precipitate was washed and suspended in water, dilute sulphuric acid was added and the mixture made up to 1 dm^3 with water. It was vigorously shaken and 25 cm^3 quickly withdrawn and this required 48.9 cm^3 of 0.002M $KMnO_4$ for oxidation. What was the percentage purity of the sample?

(23) Given that the gas constant is 0.082 atm dm^3 K^{-1} mol^{-1}, and assuming the analogy between gases and dilute solutions, find the osmotic pressure of a 3% solution of glycerol at 25°C. (Formula of glycerol is $C_3H_8O_3$.)

(24) A mixture of ethane and ethene (ethylene) was enclosed in a eudiometer tube and exploded with 100 cm^3 of oxygen. The residual volume after cooling was 61 cm^3, after absorption with potassium hydroxide 1 cm^3, and after absorption with pyrogallol zero. What was the composition of the original mixture? (All measurements were made at 15°C and atmospheric pressure.)

(25) 1.260 g of an organic compound gave on combustion 0.9786 g of carbon dioxide and 0.467 g of water. By Carius' method 1.105 g of the compound gave 1.528 g of silver iodide. 1 g of the substance dissolved in 60 g of benzene gave a depression in its freezing point of 0.49°C. Identify the compound and show how it would react with ammonia, alcoholic potassium hydroxide and magnesium. (K for 1000 g of benzene = 5.0°C.)

(26) The two oxides of titanium contain 66.7% and 60.0% of titanium respectively. The specific heat capacity of the metal is 0.544 J g^{-1} K^{-1}. Find the relative atomic mass of titanium and write formulae for its oxides.

(27) Calculate the relative molecular mass of an organic compound M from the following data:

On distilling the compound in steam the liquid boiled at 99.3°C and, of the product, 30 cm³ were water and 4.54 cm³ were M. Vapour pressure of water at 99.3°C is 98 650 N m⁻² (740 mmHg).
Atmospheric pressure was 101 300 N m⁻² (760 mmHg).
Density of M is 1.22 g cm⁻³.

(28) 25 cm³ of a solution of sodium carbonate and sodium hydroxide required 18.6 cm³ of 0.1M hydrochloric acid for the end point as determined by phenolphthalein as indicator. Another 25 cm³ of the same solution required 22.7 cm³ of 0.1M HCl, using methyl orange as indicator. Calculate the concentration of each substance in the original liquid in g dm⁻³.

(29) Calculate the enthalpy of solution of calcium chloride ($CaCl_2(s)$) from the following data:

The lattice energy of $CaCl_2(s)$ is . . -2230 kJ mol⁻¹
The enthalpy of hydration of $Ca^{2+}(g)$ is . -1640 kJ mol⁻¹
The enthalpy of hydration of $Cl^-(g)$ is . -343 kJ mol⁻¹

(30) 100 cm³ of a mixture of carbon monoxide and ethyne (acetylene) were mixed with 200 cm³ of oxygen and exploded. After cooling the gases occupied 185 cm³. Calculate the percentage by volume of each gas in the mixture. (Assume temperature and pressure constant.)

(31) Show that the Distribution Law is obeyed by the following results, obtained by shaking up iodine with a mixture of tetrachloromethane and water.

	I	II	III
g dm⁻³ of iodine in tetrachloromethane	4.98	9.28	17.1
g dm⁻³ of iodine in water layer	0.063	0.115	0.212

The temperature remained constant throughout the experiment. A certain quantity of iodine was then shaken up with water and tetrachloromethane at the same temperature, and the concentration in the tetrachloromethane layer was found to be 12.0 g dm⁻³. What would be the concentration of iodine in the aqueous layer? (University of Sheffield; adapted.)

(32) If 1 mole of ethanoic acid and ethanol are mixed together and left for some length of time, two-thirds of the alcohol are converted into the ester. Show that about 97% of the alcohol is converted into ethyl ethanoate if 10 mole of ethanoic acid are left in contact with 1 mole of ethanol. (University of Sheffield; adapted.)

(33) What conclusions do you draw from the following data:

(a) 3 g of NaCl dissolved in 1 dm³ of water had a freezing point of -0.191°C.

(b) 4 g of Na_2SO_4 dissolved in 1 dm^3 of water had a freezing point of $-0.140°C$.

(c) 6 g of urea (carbamide, CON_2H_4) per dm^3 of water froze at $-0.186°C$.

$K = 1.86°C$ per 1000 g of water.

(34) The vapour pressure of water at 13°C is 1493 N m^{-2} (11.2 mmHg). What mass of water is present in 1 dm^3 of air saturated with water vapour at 13°C and at 101 300 N m^{-2} (760 mmHg)?

(35) What will be the mass of potassium dichromate which will liberate from acidified potassium iodide the same mass of iodine as is liberated by the addition of 1 g of potassium manganate(VII)? What will be the volume of 0.142M sodium thiosulphate which will react with the iodine liberated in either case?

(36) In a Hofmann's apparatus the cross-section of the tube was 1.5 cm^2 in area, and the height of the top of the tube above the level of mercury in the trough was 120 cm.

On introducing 0.1635 g of pentanol into the tube, the level finally fell to 40 cm above the level of the trough. The temperature of the jacket was 100°C and the pressure was 101 300 N m^{-2} (760 mmHg).

Calculate the relative molecular mass of the alcohol. Assume the level of the mercury in the trough to remain constant.

(37) Calculate the percentage composition by volume of the air boiled out of water from the following data:

Air contains 79% of nitrogen by volume. } Approx.
Air contains 21% of oxygen by volume.

Air contains 0.04% of carbon dioxide by volume.

Solubility coefficients for nitrogen, oxygen and carbon dioxide are respectively 0.02, 0.04, 1.80.

(38) 25 cm^3 of a saturated solution of ethanedioic acid at room temperature were diluted to 250 cm^3 with distilled water. 25 cm^3 of the solution needed 28.0 cm^3 of 0.1M NaOH for neutralization with phenolpthalein as indicator. Calculate the solubility of ethanedioic acid as grams of the anhydrous acid per dm^3 of saturated solution under the given temperature conditions.

(39) A current of 0.1 ampere was passed through a solution of cobalt(II) and nickel(II) sulphates for 20 minutes. On the cathode was deposited an alloy containing 60% cobalt and 40% nickel by mass. What was the total mass of the alloy? (University of Sheffield.)

(40) Calculate the percentage of $Na_2B_4O_7$ in a sample of commercial borax from the observation that 4.78 g of the borax required 24.8 cm^3 1 M HCl with methyl orange as indicator.

(41) The solubilities of potassium chloride and nitrate in g of the salts per 1000 g of water are:

Temp in °C	0	10	20	30	40	50
KCl	276	310	340	370	400	426
KNO_3	133	209	316	458	639	855

Temp in °C	60	70
KCl	455	483
KNO_3	1100	1380

On the same axes and with the axis of solubility vertical, draw the graphs of the solubilities of these salts.

(a) At what temperature are the two solubilities equal?

(b) Over what temperature range is the chloride the more soluble salt?

(c) What is the solubility of the nitrate at 25°C and 55°C?

(d) What mass of crystals would be deposited by cooling a saturated solution of the nitrate, containing 30 g of water, from 70°C, at which it was just saturated, to 10°C? What percentage is this of the mass of salt originally dissolved?

(e) If the same as in (d) is done for the chloride what are the corresponding figures? Which is the more suitable salt for this type of crystallization?

(f) At what temperatures do these salts differ in solubility by 50 g?

(g) In the light of Le Chatelier's Principle what can you deduce from the graphs about the enthalpies of solution of the two salts?

(42) A certain compound, X, shows the following figures for solubility in g per 1000 g of water:

Temp. in °C	0	10	20	30	32.5
	50	90	194	408	465

Starting from a derivative of X, the following figures for solubility can also be obtained:

Temp. in °C	40	50	60	70	80
	488	467	453	445	437

On the same axes and with the axis of solubility vertical, draw the graphs from the data given.

(a) State briefly what other data you would consider desirable to complete the picture of the solubility relations of X.

(b) By extrapolation of the two graphs, obtain an approximate value for the temperature at which the solubility values coincide.

(c) Given that X is a salt hydrate, can you make a suggestion to explain the unusual solubility behaviour which it shows?

(d) By reference to Le Chatelier's Principle compare the enthalpies of solution of X over the range 0–30°C and of its derivative over the higher temperature range.

(e) To a saturated solution containing 1000 g of water, and saturated with the derivative of X at 80°C, are added a further 50 g of X. The mixture is then subjected to very slow cooling and stirring till it reaches 10°C. What would you expect to see happen in the mixture in the course of this cooling? Give your reasons.

(43) 0.01 g of lead(II) iodide, containing a small quantity of the radioactive isotope ^{131}I, was stirred with about 10 cm^3 of distilled water at a constant temperature and then centrifuged. 1 cm^3 of the supernatant liquid on evaporation gave a residue which had an activity $\frac{1}{15}$ of that of the original 0.01 g of lead iodide. Calculate the solubility of lead(II) iodide in g per 1000 g of water and its solubility product at this temperature.

Answers to Numerical Examples

The answers have been calculated using the relative atomic masses on page 132.

CHAPTER 1, p. 1

(1) 340.5
(2) 31.08
(3) 5.212
    57.82
(4) 0.9061
(5) 0.7571
(6) 4.971
(7) 14.10
    13.52
(8) 57.6
(9) 12.93
(10) 1.052
     3.948

(11) 2.88
     2.12
(12) 96.6
(13) 10.8
(14) 0.292
(15) 26.64 cm^3
(16) 98.37
(17) 31.5
     68.5
(18) 31.82
(19) 10
(20) 2
(21) 2

(22) 5
(23) 28 g
     56
     40
(24) 1
     Z^{2+}
(25) 64.9 g
     129.8
(26) 100
     40
(27) 45 g
     2

CHAPTER 2, p. 12

(1) 0.0203
(2) 0.0206
(3) 0.0194
    3.063
(4) 6.04
(5) 99.45
(6) 7
(7) 11.6
(8) 4.166
    3.583
(9) 13.98
    13.60
(10) 3.168
     5.037
(11) 114.2
(12) 39.93

(13) 1.65
(14) 32.5
     10.7
(15) 2
(16) 75.82
(17) 1.95
(18) 3:5
(19) 1:2
(20) 6
(21) 85
(22) 0.1033
(23) 25.1
     12 cm^3
(24) 12.98 g
(25) 0.009 527
(26) 50.58

(27) 0.4327
(28) 32.5
(29) 74.30
(30) 0.7491 g
     5.498 g
(31) 5.937 g
(32) 8.602 g
(33) 3.48
(34) 2:3
(35) 3:1
     C$_6$H$_2$Br$_3$NH$_2$

CHAPTER 3, p. 27

(1) 65.31
(2) 95.9

(3) 3.41
(4) 83.92

(5) 2.657
    6.616

(6) 46.68
 53.32
(7) 2
(8) 68.98

(9) 60
 2.96
(10) 46.00
(11) 28.95

(12) 7:8:10
(13) PCl_5
(14) 110.5 g
 39.5

Chapter 4, p. 32

(1) 56.14
(2) 0.323 g
(3) 43.94
 56.06

(4) 35.67
(5) 14.76
(6) 64
(7) 7.0

(8) 62.9
(9) 5

Chapter 5, p. 36

(1) 55.7
(2) -285 N m^{-2}
(3) 179
(4) 1.86°C
 per 1000 g
(5) 0.3453 g
(6) -0.8266°C
(7) 80.568°C
(8) 60
(9) 59
(10) 155.2
(11) 252
(12) 59
(13) 92

(14) 86.84
(15) 121.8

(16) $M_r = 125.3$
 $\therefore P_4$
(17) $M_r = 251.9$
 $\therefore S_8$
(18) 915 g
(19) 94 800 N m^{-2}
(20) 19.35 g
(21) 342
(22) 18.44
(23) 180
 0.491 atm
(24) 872 000 N m^{-2}
(25) 61.69
 -0.03°C

(26) 92.23
(27) 342

(28) 0.0974 g
(29) 10.83 g
(30) $\dfrac{\text{glu}}{\text{gly}} = \dfrac{45}{23}$
 $= 1.957$
(31) 1.76
(32) $R = 0.082$
 (gas)
 0.083
 0.079
 0.081
 0.080
 $\overline{0.081}$ Mean
(33) 7.97
(34) 117 300

Chapter 6, p. 49

(1) 0.939
(2) 3996 N m^{-2}
(3) 86
(4) -0.2127°C
(5) 80.4%
(6) 0.72
(7) 46 530

(8) 74%
(9) 80.9%
(10) 0.145
(11) 0.3795 g
(12) 75.25%
(13) 93.2%
(14) 841 300

(15) 61.95 g
(16) 90.3%
(17) Dimer (approx.)
(18) Dimer (approx.)
(19) 4 ions from
 each salt
(20) -0.36°C

Chapter 7, p. 54

(1) 43.94
(2) 31.98
(3) 36.85
 73.7
(4) 44.16
(5) 73.9

(6) 59.8
 119.6
(7) 116.1
(8) 180.5
(9) 79.29
(10) $M_r(H_2O) = 17.72$

(11) 16
 32
 32
 48 $\therefore$ 16
 16
 16

(12) 12
    72
    36
    12 ∴ 12
    12
    12
    24
(13) 64

32 ∴ 32
32
32
(14) 23.74
(15) 206.4
     206.88
     Mean 206.64
     MO

MO_2
(16) 2
    9.05
(17) 55.0
(18) 79.0
(19) 28
(20) 246 s

CHAPTER 8, p. 63

(1) 0.8448
(2) 80.8
(3) 10.16
(4) 62.8
(5) HI 77.0
    H_2 0.180
    I_2 22.82

(6) % diss.
    30
    41
    52
    63
    74
    1560°C

(7) 12.891 cm^3
(8) 22.73 cm^3
    0.00375 g
(9) 1.969 dm^3
    7.275 dm^3
(10) 0.28
(11) 0.61

CHAPTER 9, p. 67 (All answers here in kJ mol^{-1})

(1) −246
(2) −71
(3) −727
(4) −110
(5) +90
(6) −2726
(7) −29
(8) +216.3
(9) −126

(10) −210
(11) −322
(12) +130
(13) −77.7
(14) −57.3
(15) −57.1
(16) −58.2
(17) −54.6
(18) −780

(19) −2250
     −590
(20) −130
(21) −3220
     −234
(22) −390
(23) −703
(24) −2176

CHAPTER 10, p. 77

(1) 5.33
    5.46
    5.22
    5.20
    5.29
    Mean 5.30
(2) I. 87.94
    II. 88.33

III. 87.50
     Mean 87.92
     0.1112 g
(3) 0.44 g
    0.14 g
(4) 86.4
(5) 11
    4

2

(6) $\dfrac{\sqrt{C_B}}{C_W} = 0.746$

(7) $\dfrac{\sqrt{C_B}}{C_W} = 33.5$

(8) 1.25 kg.

CHAPTER 11, p. 82

(1) 0.000 33 g C^{-1}
(2) 9.07 g
(3) 0.0000103 g C^{-1}
    0.0000832 g C^{-1}

0.0000936 g C^{-1}
0.000338 g C^{-1}
(4) $\dfrac{4xy}{3}$

(5) 111 cm^3
(6) 0.709 g
    79.6 cm^3
(7) 0.493 g

Chapter 12, p. 85

(1) 4
   (a) 0.42 ⎫ moles
   (b) 0.97 ⎬ of
   (c) 0.54 ⎭ ester
(2) H_2 0.6
   I_2 40.6
   HI 38.8
(3) 58.7
(4) 27 760 N m^{-2}
   55 540 N m^{-2}
(5) 5.3 atm
(6) 1.16
(7) (a) 27.9 CO, H_2O
      22.1 CO_2, H_2
   (b) 14.6 H_2O
      39.6 CO
      35.4 H_2
      10.4 CO_2
(10) 6.82×10^{-3} M
    11.8

(11) 1.99×10^{-5} M
    4.70
    4.46×10^{-2}
    4.46×10^{-4} M
    10.6
(12) 2
    11
    1.26
    12.7
    2.85
(13) (a) 1.48
    (b) 3
    (c) 11
    (d) 11.6
    (e) 1.7
(14) 0.027
    7.5×10^{-4}
    mol dm^{-3}
    3.12
    3.2

(16) 1.4×10^{-3} to
    1.07×10^{-5} mol dm^{-3}
(17) $-0.0392°C$
(18) 2.48
(19) 0.046
    0.08
(21) 0.0083 g
(22) 1×10^{-10} mol^2 dm^{-6}
    1.44×10^{-6} g dm^{-3}
(23) 10^{-2}
(24) 4.66×10^{-4}
(25) 1.19×10^{-4}
    8.95×10^{-10}
(26) 16.74 g
(27) No
    $-4.8 \times 10^{-5}°C$
(28) 1.36×10^{-2} g
    7.2×10^{-6} g
(29) 0.0363 g

Chapter 13, p. 100

(1) C_2H_2
(2) C_4H_{10}
(3) C_2H_6
(4) C_2H_4
(5) C_3H_4
(6) C_3H_8

Chapter 14, p. 106

(1) CH_3COOH
(2) C_2H_6O
(3) C_6H_6
(4) $C_3H_6O_2$
(5) C_7H_8
(6) CH_3OH
(7) C_6H_7N
(8) $(CH_3)_2CO$
(9) CH_3CONH_2
(10) $C_2H_5NH_2$
(11) C_6H_5COOH
(12) 90
    $H_2C_2O_4$
(13) $(CH_3)_3N$
(14) C_2H_5CHO
(15) $C_6H_5N(CH_3)_2$
(16) CH_3CHO
(17) C_2H_5CN
    $C_2H_5CH_2NH_2$
(18) C_2H_5Br
(19) $C_6H_5CH_2Cl$
    $CH_3C_6H_4Cl$
(20) CH_3CHCl_2
    $(CH_2Cl)_2$
(21) $1,3-C_6H_4(NO_2)_2$
(22) C_2H_5OH
    $(CH_3)_2O$
(23) $C_6H_5CH_2NH_2$
    $CH_3C_6H_4NH_2$
(24) $(CH_3)(C_2H_5)CO$
(25) C_3H_9N
(26) $CH_3CCl_2CH_3$
(27) CH_3CH_2CHO
    $(CH_3)_2CO$
(28) $CHCl_3$
(29) $C_6H_5CH_2OH$
(30) C_5H_{12}
(31) $CH_3CH_2CH_2OH$
    $CH_3CH(OH)CH_3$
(32) $C_6H_5CCl_3$
(33) C_6H_5OH
(34) $C_2H_5OCH_3$
(35) C_6H_5CHO
(36) $C_6H_4(SO_3H)_2$
(37) $1,2-;\ 1,3-;$
    and $1,4-;$
    $C_6H_4(NH_2)(NO_2)$
(38) G.A.S.L.A.

Chapter 15, p. 111

(1) 21.04
(2) 151.25 cm^3
(3) 46.1
    46.1
    7.7
(4) 17 cm^3
(5) 10 cm^3
    5 cm^3
    45 cm^3
(6) 43.3
    56.7

Chapter 16, p. 114

(1) 14.46
(2) 38.24
    31.87
    15.54
    ∴ NH$_4$ 14.35
    KBr
    (NH$_4$)$_2$SO$_4$
(3) CuSO$_4$.5H$_2$O
    CuSO$_4$.H$_2$O
    CuSO$_4$
    CuO
(4) 0.1355 dm^3
(5) 19.27
    80.00
    0.7254
(6) 34.76
(7) 74
(8) O$_2$ 56.8
    CO$_2$ 40.0
    N$_2$ 3.2
(9) 51.23
(10) 1.292 atm
     3.461 g dm^{-3}
(11) 0.04239
(12) 96.34
(13) $\dfrac{49}{54} = 0.9074$
(14) 60
     156.8
(15) 337.9 g
(16) O$_2$ 49.77
     CO$_2$ 40.64
     I.G. 9.59
(17) 26.54
     12
(18) C$_6$H$_5$SO$_3$H
(19) 0.0003284 g C^{-1}
     6.983 g
     754.0 cm^3
(20) 7.09
(21) K = 1.85°C/kg H$_2$O
     −0.308°C
(22) 97.8
(23) 7.967 atm
(24) 18 cm^3
     12 cm^3
(25) C$_3$H$_7$I
(26) 48.04
     Ti$_2$O$_3$
     TiO$_2$
(27) 123
(28) 1.739
     2.320
(29) −96 kJ mol^{-1}
(30) 35
     65
(31) Av. 80.2
     0.1496 g dm^{-3}
(32) 97.3%
(33) $i = 2$
     $i = 2.67$
     $i = 1$
(34) 0.01131 g
(35) 1.551 g
     222.8 cm^3
(36) 88
(37) 63.40
     33.70
     2.89
(38) 50.4
(39) 0.03657 g
(40) 52.4
(41) 22°C
     0–22°C
     380 g; 970 g
     35.1 g; 84.8
     5.2 g; 35.8
     8°C; 31°C
(42) 34°C
(43) 6.67
     1.21 × 10^{-8} mol^3 dm^{-9}

LOGARITHMS

	0	1	2	3	4	5	6	7	8	9	1	2	3	4	5	6	7	8	9
10	0000	0043	0086	0128	0170	0212	0253	0294	0334	0374	4	8	12	17	21	25	29	33	37
11	0414	0453	0492	0531	0569	0607	0645	0682	0719	0755	4	8	11	15	19	23	26	30	34
12	0792	0828	0864	0899	0934	0969	1004	1038	1072	1106	3	7	10	14	17	21	24	28	31
13	1139	1173	1206	1239	1271	1303	1335	1367	1399	1430	3	6	10	13	16	19	23	26	29
14	1461	1492	1523	1553	1584	1614	1644	1673	1703	1732	3	6	9	12	15	18	21	24	27
15	1761	1790	1818	1847	1875	1903	1931	1959	1987	2014	3	6	8	11	14	17	20	22	25
16	2041	2068	2095	2122	2148	2175	2201	2227	2253	2279	3	5	8	11	13	16	18	21	24
17	2304	2330	2355	2380	2405	2430	2455	2480	2504	2529	2	5	7	10	12	15	17	20	22
18	2553	2577	2601	2625	2648	2672	2695	2718	2742	2765	2	5	7	9	12	14	16	19	21
19	2788	2810	2833	2856	2878	2900	2923	2945	2967	2989	2	4	7	9	11	13	16	18	20
20	3010	3032	3054	3075	3096	3118	3139	3160	3181	3201	2	4	6	8	11	13	15	17	19
21	3222	3243	3263	3284	3304	3324	3345	3365	3385	3404	2	4	6	8	10	12	14	16	18
22	3424	3444	3464	3483	3502	3522	3541	3560	3579	3598	2	4	6	8	10	12	14	15	17
23	3617	3636	3655	3674	3692	3711	3729	3747	3766	3784	2	4	6	7	9	11	13	15	17
24	3802	3820	3838	3856	3874	3892	3909	3927	3945	3962	2	4	5	7	9	11	12	14	16
25	3979	3997	4014	4031	4048	4065	4082	4099	4116	4133	2	3	5	7	9	10	12	14	15
26	4150	4166	4183	4200	4216	4232	4249	4265	4281	4298	2	3	5	7	8	10	11	13	15
27	4314	4330	4346	4362	4378	4393	4409	4425	4440	4456	2	3	5	6	8	9	11	13	14
28	4472	4487	4502	4518	4533	4548	4564	4579	4594	4609	2	3	5	6	8	9	11	12	14
29	4624	4639	4654	4669	4683	4698	4713	4728	4742	4757	1	3	4	6	7	9	10	12	13
30	4771	4786	4800	4814	4829	4843	4857	4871	4886	4900	1	3	4	6	7	9	10	11	13
31	4914	4928	4942	4955	4969	4983	4997	5011	5024	5038	1	3	4	6	7	8	10	11	12
32	5051	5065	5079	5092	5105	5119	5132	5145	5159	5172	1	3	4	5	7	8	9	11	12
33	5185	5198	5211	5224	5237	5250	5263	5276	5289	5302	1	3	4	5	6	8	9	10	12
34	5315	5328	5340	5353	5366	5378	5391	5403	5416	5428	1	3	4	5	6	8	9	10	11
35	5441	5453	5465	5478	5490	5502	5514	5527	5539	5551	1	2	4	5	6	7	9	10	11
36	5563	5575	5587	5599	5611	5623	5635	5647	5658	5670	1	2	4	5	6	7	8	10	11
37	5682	5694	5705	5717	5729	5740	5752	5763	5775	5786	1	2	3	5	6	7	8	9	10
38	5798	5809	5821	5832	5843	5855	5866	5877	5888	5899	1	2	3	5	6	7	8	9	10
39	5911	5922	5933	5944	5955	5966	5977	5988	5999	6010	1	2	3	4	5	7	8	9	10
40	6021	6031	6042	6053	6064	6075	6085	6096	6107	6117	1	2	3	4	5	6	8	9	10
41	6128	6138	6149	6160	6170	6180	6191	6201	6212	6222	1	2	3	4	5	6	7	8	9
42	6232	6243	6253	6263	6274	6284	6294	6304	6314	6325	1	2	3	4	5	6	7	8	9
43	6335	6345	6355	6365	6375	6385	6395	6405	6415	6425	1	2	3	4	5	6	7	8	9
44	6435	6444	6454	6464	6474	6484	6493	6503	6513	6522	1	2	3	4	5	6	7	8	9
45	6532	6542	6551	6561	6571	6580	6590	6599	6609	6618	1	2	3	4	5	6	7	8	9
46	6628	6637	6646	6656	6665	6675	6684	6693	6702	6712	1	2	3	4	5	6	7	7	8
47	6721	6730	6739	6749	6758	6767	6776	6785	6794	6803	1	2	3	4	5	5	6	7	8
48	6812	6821	6830	6839	6848	6857	6866	6875	6884	6893	1	2	3	4	4	5	6	7	8
49	6902	6911	6920	6928	6937	6946	6955	6964	6972	6981	1	2	3	4	4	5	6	7	8
50	6990	6998	7007	7016	7024	7033	7042	7050	7059	7067	1	2	3	3	4	5	6	7	8
51	7076	7084	7093	7101	7110	7118	7126	7135	7143	7152	1	2	3	3	4	5	6	7	8
52	7160	7168	7177	7185	7193	7202	7210	7218	7226	7235	1	2	2	3	4	5	6	7	7
53	7243	7251	7259	7267	7275	7284	7292	7300	7308	7316	1	2	2	3	4	5	6	6	7
54	7324	7332	7340	7348	7356	7364	7372	7380	7388	7396	1	2	2	3	4	5	6	6	7
	0	1	2	3	4	5	6	7	8	9	1	2	3	4	5	6	7	8	9

LOGARITHMS

	0	1	2	3	4	5	6	7	8	9	1	2	3	4	5	6	7	8	9
55	7404	7412	7419	7427	7435	7443	7451	7459	7466	7474	1	2	2	3	4	5	5	6	7
56	7482	7490	7497	7505	7513	7520	7528	7536	7543	7551	1	2	2	3	4	5	5	6	7
57	7559	7566	7574	7582	7589	7597	7604	7612	7619	7627	1	2	2	3	4	5	5	6	7
58	7634	7642	7649	7657	7664	7672	7679	7686	7694	7701	1	1	2	3	4	4	5	6	7
59	7709	7716	7723	7731	7738	7745	7752	7760	7767	7774	1	1	2	3	4	4	5	6	7
60	7782	7789	7796	7803	7810	7818	7825	7832	7839	7846	1	1	2	3	4	4	5	6	6
61	7853	7860	7868	7875	7882	7889	7896	7903	7910	7917	1	1	2	3	4	4	5	6	6
62	7924	7931	7938	7945	7952	7959	7966	7973	7980	7987	1	1	2	3	3	4	5	6	6
63	7993	8000	8007	8014	8021	8028	8035	8041	8048	8055	1	1	2	3	3	4	5	5	6
64	8062	8069	8075	8082	8089	8096	8102	8109	8116	8122	1	1	2	3	3	4	5	5	6
65	8129	8136	8142	8149	8156	8162	8169	8176	8182	8189	1	1	2	3	3	4	5	5	6
66	8195	8202	8209	8215	8222	8228	8235	8241	8248	8254	1	1	2	3	3	4	5	5	6
67	8261	8267	8274	8280	8287	8293	8299	8306	8312	8319	1	1	2	3	3	4	5	5	6
68	8325	8331	8338	8344	8351	8357	8363	8370	8376	8382	1	1	2	3	3	4	4	5	6
69	8388	8395	8401	8407	8414	8420	8426	8432	8439	8445	1	1	2	2	3	4	4	5	6
70	8451	8457	8463	8470	8476	8482	8488	8494	8500	8506	1	1	2	2	3	4	4	5	6
71	8513	8519	8525	8531	8537	8543	8549	8555	8561	8567	1	1	2	2	3	4	4	5	5
72	8573	8579	8585	8591	8597	8603	8609	8615	8621	8627	1	1	2	2	3	4	4	5	5
73	8633	8639	8645	8651	8657	8663	8669	8675	8681	8686	1	1	2	2	3	4	4	5	5
74	8692	8698	8704	8710	8716	8722	8727	8733	8739	8745	1	1	2	2	3	4	4	5	5
75	8751	8756	8762	8768	8774	8779	8785	8791	8797	8802	1	1	2	2	3	3	4	5	5
76	8808	8814	8820	8825	8831	8837	8842	8848	8854	8859	1	1	2	2	3	3	4	5	5
77	8865	8871	8876	8882	8887	8893	8899	8904	8910	8915	1	1	2	2	3	3	4	4	5
78	8921	8927	8932	8938	8943	8949	8954	8960	8965	8971	1	1	2	2	3	3	4	4	5
79	8976	8982	8987	8993	8998	9004	9009	9015	9020	9025	1	1	2	2	3	3	4	4	5
80	9031	9036	9042	9047	9053	9058	9063	9069	9074	9079	1	1	2	2	3	3	4	4	5
81	9085	9090	9096	9101	9106	9112	9117	9122	9128	9133	1	1	2	2	3	3	4	4	5
82	9138	9143	9149	9154	9159	9165	9170	9175	9180	9186	1	1	2	2	3	3	4	4	5
83	9191	9196	9201	9206	9212	9217	9222	9227	9232	9238	1	1	2	2	3	3	4	4	5
84	9243	9248	9253	9258	9263	9269	9274	9279	9284	9289	1	1	2	2	3	3	4	4	5
85	9294	9299	9304	9309	9315	9320	9325	9330	9335	9340	1	1	2	2	3	3	4	4	5
86	9345	9350	9355	9360	9365	9370	9375	9380	9385	9390	1	1	2	2	3	3	4	4	5
87	9395	9400	9405	9410	9415	9420	9425	9430	9435	9440	0	1	1	2	2	3	3	4	4
88	9445	9450	9455	9460	9465	9469	9474	9479	9484	9489	0	1	1	2	2	3	3	4	4
89	9494	9499	9504	9509	9513	9518	9523	9528	9533	9538	0	1	1	2	2	3	3	4	4
90	9542	9547	9552	9557	9562	9566	9571	9576	9581	9586	0	1	1	2	2	3	3	4	4
91	9590	9595	9600	9605	9609	9614	9619	9624	9628	9633	0	1	1	2	2	3	3	4	4
92	9638	9643	9647	9652	9657	9661	9666	9671	9675	9680	0	1	1	2	2	3	3	4	4
93	9685	9689	9694	9699	9703	9708	9713	9717	9722	9727	0	1	1	2	2	3	3	4	4
94	9731	9736	9741	9745	9750	9754	9759	9763	9768	9773	0	1	1	2	2	3	3	4	4
95	9777	9782	9786	9791	9795	9800	9805	9809	9814	9818	0	1	1	2	2	3	3	4	4
96	9823	9827	9832	9836	9841	9845	9850	9854	9859	9863	0	1	1	2	2	3	3	4	4
97	9868	9872	9877	9881	9886	9890	9894	9899	9903	9908	0	1	1	2	2	3	3	4	4
98	9912	9917	9921	9926	9930	9934	9939	9943	9948	9952	0	1	1	2	2	3	3	4	4
99	9956	9961	9965	9969	9974	9978	9983	9987	9991	9996	0	1	1	2	2	3	3	3	4
	0	1	2	3	4	5	6	7	8	9	1	2	3	4	5	6	7	8	9

Relative Atomic Masses

Aluminium	27	Manganese	55
Barium	137	Mercury	201
Boron	11	Nitrogen	14
Bromine	80	Oxygen	16
Calcium	40	Phosphorus	31
Carbon	12	Platinum	195
Chlorine	35.5	Potassium	39
Chromium	52	Silver	108
Copper	64	Sodium	23
Hydrogen	1	Strontium	88
Iodine	127	Sulphur	32
Iron	56	Tin	119
Lead	207	Zinc	65
Magnesium	24		

Other Useful Data

The molar volume of a gas, V_m, is $22.4\,dm^3$ at s.t.p.
The Faraday constant, F, is $96\,500\,C\,mol^{-1}$.
The density of air is $1.293\,g\,dm^{-3}$ at s.t.p.
The density of hydrogen is $0.09\,g\,dm^{-3}$ at s.t.p.
The density of water is $1\,kg\,dm^{-3}$.
1 metric ton (tonne) is $1000\,kg$.
$1\,dm^3$ (litre) is $1000\,cm^3$.